HISTORY, DEVELOPMENT AND FUTURE OF SLOWPOKE AND MNSR RESEARCH REACTORS

The following States are Members of the International Atomic Energy Agency:

AFGHANISTAN
ALBANIA
ALGERIA
ANGOLA
ANTIGUA AND BARBUDA
ARGENTINA
ARMENIA
AUSTRALIA
AUSTRIA
AZERBAIJAN
BAHAMAS, THE
BAHRAIN
BANGLADESH
BARBADOS
BELARUS
BELGIUM
BELIZE
BENIN
BOLIVIA, PLURINATIONAL
  STATE OF
BOSNIA AND HERZEGOVINA
BOTSWANA
BRAZIL
BRUNEI DARUSSALAM
BULGARIA
BURKINA FASO
BURUNDI
CABO VERDE
CAMBODIA
CAMEROON
CANADA
CENTRAL AFRICAN
  REPUBLIC
CHAD
CHILE
CHINA
COLOMBIA
COMOROS
CONGO
COOK ISLANDS
COSTA RICA
CÔTE D'IVOIRE
CROATIA
CUBA
CYPRUS
CZECH REPUBLIC
DEMOCRATIC REPUBLIC
  OF THE CONGO
DENMARK
DJIBOUTI
DOMINICA
DOMINICAN REPUBLIC
ECUADOR
EGYPT
EL SALVADOR
ERITREA
ESTONIA
ESWATINI
ETHIOPIA
FIJI
FINLAND
FRANCE
GABON
GAMBIA, THE

GEORGIA
GERMANY
GHANA
GREECE
GRENADA
GUATEMALA
GUINEA
GUYANA
HAITI
HOLY SEE
HONDURAS
HUNGARY
ICELAND
INDIA
INDONESIA
IRAN, ISLAMIC REPUBLIC OF
IRAQ
IRELAND
ISRAEL
ITALY
JAMAICA
JAPAN
JORDAN
KAZAKHSTAN
KENYA
KOREA, REPUBLIC OF
KUWAIT
KYRGYZSTAN
LAO PEOPLE'S DEMOCRATIC
  REPUBLIC
LATVIA
LEBANON
LESOTHO
LIBERIA
LIBYA
LIECHTENSTEIN
LITHUANIA
LUXEMBOURG
MADAGASCAR
MALAWI
MALAYSIA
MALI
MALTA
MARSHALL ISLANDS
MAURITANIA
MAURITIUS
MEXICO
MONACO
MONGOLIA
MONTENEGRO
MOROCCO
MOZAMBIQUE
MYANMAR
NAMIBIA
NEPAL
NETHERLANDS,
  KINGDOM OF THE
NEW ZEALAND
NICARAGUA
NIGER
NIGERIA
NORTH MACEDONIA
NORWAY
OMAN

PAKISTAN
PALAU
PANAMA
PAPUA NEW GUINEA
PARAGUAY
PERU
PHILIPPINES
POLAND
PORTUGAL
QATAR
REPUBLIC OF MOLDOVA
ROMANIA
RUSSIAN FEDERATION
RWANDA
SAINT KITTS AND NEVIS
SAINT LUCIA
SAINT VINCENT AND
  THE GRENADINES
SAMOA
SAN MARINO
SAUDI ARABIA
SENEGAL
SERBIA
SEYCHELLES
SIERRA LEONE
SINGAPORE
SLOVAKIA
SLOVENIA
SOMALIA
SOUTH AFRICA
SPAIN
SRI LANKA
SUDAN
SWEDEN
SWITZERLAND
SYRIAN ARAB REPUBLIC
TAJIKISTAN
THAILAND
TOGO
TONGA
TRINIDAD AND TOBAGO
TUNISIA
TÜRKİYE
TURKMENISTAN
UGANDA
UKRAINE
UNITED ARAB EMIRATES
UNITED KINGDOM OF
  GREAT BRITAIN AND
  NORTHERN IRELAND
UNITED REPUBLIC OF TANZANIA
UNITED STATES OF AMERICA
URUGUAY
UZBEKISTAN
VANUATU
VENEZUELA, BOLIVARIAN
  REPUBLIC OF
VIET NAM
YEMEN
ZAMBIA
ZIMBABWE

TECHNICAL REPORTS SERIES No. 497

HISTORY, DEVELOPMENT AND FUTURE OF SLOWPOKE AND MNSR RESEARCH REACTORS

INTERNATIONAL ATOMIC ENERGY AGENCY
VIENNA, 2025

COPYRIGHT NOTICE

© IAEA, 2025

Printed by the IAEA in Austria
November 2025
STI/DOC/010/497
https://doi.org/10.61092/iaea.nj1m-x1cm

IAEA Library Cataloguing in Publication Data

Names: International Atomic Energy Agency.
Title: History, development and future of SLOWPOKE and MNSR research reactors / International Atomic Energy Agency.
Description: Vienna : International Atomic Energy Agency, 2025. | Series: Technical reports series, ISSN 0074–1914 ; no. 497 | Includes bibliographical references.
Identifiers: IAEAL 25-01763 | ISBN 978-92-0-109425-4 (paperback : alk. paper) | ISBN 978-92-0-109525-1 (pdf) | ISBN 978-92-0-109625-8 (epub)
Subjects: LCSH: Nuclear reactors — History. | Nuclear reactors — Research. Nuclear reactors — Management. | Nuclear reactors — Design and construction.
Classification: UDC 621.039.57 | STI/DOC/010/497

FOREWORD

The prototype Safe Low Power Critical Experiment research reactor (known as SLOWPOKE-1) was commissioned in 1970 in Canada, followed by SLOWPOKE-2 in 1971. The miniature neutron source reactor (MNSR) prototype was commissioned in 1984 in China and the first MNSR outside China was commissioned in Pakistan in 1989. In total, 12 SLOWPOKE reactors and 10 MNSRs have been built and operated, and 11 still operate today in 8 countries. These facilities have played an important role in these countries by providing neutron activation analysis, research, education and training opportunities, radiography capability and limited isotope production for many decades. Key design objectives were to avoid the high operating costs of larger research reactors and to provide a core lifetime with low fuel consumption, a low number of operational staff required and a comparatively simple reactivity control system. None of the reactors have been closed down because of ageing or failure of a major component. The only original major system that has been changed in most of the reactors is the analogue control system. None of the reactors have required any long outage because of the failure of a major component. Nine of these reactors have already been decommissioned, without any significant challenges or issues.

The motivation for this publication was the release of Technical Reports Series No. 482, History, Development and Future of TRIGA Research Reactors, and a consultancy meeting held in 2021. During the meeting, participants recognized the value in producing a similar publication focused on SLOWPOKE reactors and MNSRs and agreed to contribute to the preparation of this publication.

This publication offers a source of reference information on these research reactors, covering aspects such as their history, development, utilization, operational experience, core conversion and refuelling, ageing management and decommissioning, and addressing issues, challenges and future opportunities associated with these reactors. In addition, the Annex provides the abstracts of 17 papers illustrating the historical and operational experience and possible future development of SLOWPOKE reactors and MNSRs under construction, in operation, or shut down and decommissioned.

The IAEA is grateful to all the individual contributors who provided input and reviewed this publication, in particular D.J. Winfield (Canada) and Y. Li (China). The IAEA officers responsible for this publication were N. Pessoa Barradas and M.-K. Gavello of the Division of Physical and Chemical Sciences.

CONTENTS

1. INTRODUCTION

1.1. BACKGROUND

The Safe Low Power Critical Experiment (SLOWPOKE) reactors and miniature neutron source reactors (MNSRs) were first developed in the late 1970s. A total of 22 have been built to date in 8 different countries, with 11 now decommissioned after successful operation. One is under construction in a ninth country. All reactors are listed in Section 2.[1] The oldest has been operating since 1976. The original design lifetime was expected to be about 20 to 25 years. The reactors were not originally designed to be refuelled; however, eight reactors have been successfully refuelled to date. In several cases, the main goal of the refuelling was conversion from high enriched uranium (HEU) to low enriched uranium (LEU).

The objective for the design of the SLOWPOKE reactor was to produce a neutron source with a high degree of reactivity safety and a very small fuel core, while having a neutron flux as high as feasible for neutron activation analysis (NAA). Other prime considerations were to reduce the cost, design and operational complexity and to avoid the higher staffing levels needed for the Training, Research, Isotopes, General Atomics (TRIGA) reactors that were widely available at the time.

1.2. OBJECTIVE

The objective of this publication is to summarize the information on the history, design, safety, utilization and operation of SLOWPOKE reactors and MNSRs. The publication also provides a view of potential future challenges that need to be addressed by current operating organizations. Guidance and recommendations provided here in relation to identified good practices represent expert opinion but are not made on the basis of a consensus of all Member States.

1.3. SCOPE

This publication focuses on the history, design, technical characteristics, operational information, safety, applications, issues and challenges of

---

[1] All research reactor names in this publication are given as they appear in the IAEA's Research Reactor Database (RRDB), https://nucleus.iaea.org/rrdb/

SLOWPOKE and MNSR research reactors. It includes information on reactors that are in operation, decommissioned and under construction, and on associated projects that were never built.

1.4. STRUCTURE

This publication consists of nine sections. The Introduction is followed by a description of the historical development of SLOWPOKEs and MNSRs (Section 2), basic reactor characteristics (Sections 2 and 3), applications (Section 4), LEU fuel conversion and refuelling (Section 5), operational experience (Section 6), ageing management (Section 7), decommissioning (Section 8) and future opportunities (Section 9). The Annex to this publication includes the abstracts of 17 facility reports from Member States describing many of the SLOWPOKE and MNSR facilities worldwide.

2. HISTORY

2.1. HISTORICAL BASIS

During the 1950s and 1960s, a variety of organizations worked on international projects to design and construct research reactors. These were predominantly neutron production facilities used for research and training, fuel and materials testing, and production of radioisotopes for medical and industrial use. Research reactors larger than 10 MW(th) were mainly operated by Member State national nuclear authorities. Research reactor development in the 1970s progressed towards higher neutron fluxes, culminating in 1971 with the Laue-Langevin Institute reactor, Grenoble, with the highest continuous thermal neutron flux available in the world at that time (1.5×10^{15} cm$^{-2} \cdot$s^{-1}, with a thermal power of 58.3 MW), which was exclusively used for neutron scattering and related research. Such reactors were complex and costly, requiring a significant number of full time operational and support staff.

In the mid-1950s, the TRIGA series of reactors was developed, with the first TRIGA Mark I reaching criticality in 1958 [1]. Describing the dominant design feature for mitigating reactivity insertion accidents, one of the original designers, Edward Teller, stated that he and his colleagues should "design a reactor so safe that it could be given to a bunch of high school children to play with, without any fear that they would get hurt" [2]. Key to the TRIGA design concept was the use of a unique uranium zirconium hydride (UZrH) fuel design to achieve a high degree of safety for reactivity insertions, along with the ability to operate in pulse mode. The large prompt and inherent negative temperature feedback reduced the reliance upon the engineered safety systems that were required for all research reactor types at that time.

The first TRIGA reactors were mainly suitable for universities and were kept to low power levels, in the 10–100 kW range. By June 1967, 32 TRIGA facilities had been installed in 13 countries, with steady state powers ranging from 250 kW to 4 MW, subsequently culminating in 1980 with a maximum of 14 MW power [1]. In comparison, the SLOWPOKE concept did not originate until 1967. It arose from an internal Los Alamos National Laboratory experimental paper distributed by Jarvis and Mills in March 1967 [3]. This internal paper announced that a new low value for the critical mass of ^{235}U in a critical reactor assembly had been confirmed: 250 g of 93% ^{235}U in a homogeneous polyethylene core surrounded by a thick beryllium reflector. In June 1967 the announcement reached the wider nuclear community [4, 5]. In September 1967, J.W. Hilborn — then at Atomic Energy of Canada Limited (AECL), Whiteshell Laboratories — informed by this announcement, recognized the possibility of designing and constructing a simple, low cost subcritical fission neutron source facility. Such a facility was intended for universities, mainly for NAA use, which was increasing rapidly at that time. A key design objective from Hilborn's proposal was to achieve a neutron flux with a factor of ten greater than was then available ($\approx 10^{11}$ cm$^{-2}\cdot$s^{-1}) from existing accelerator based neutron generators. The subcritical fission source would be simpler and more reliable and would produce a more continuous neutron source for around the same price as an accelerator based neutron generator. The classic version of accelerator based neutron generators at that time used the ^{3}H(d,n)^{4}He reaction. This had a low neutron flux, a limited lifetime, and reliability and maintenance issues. Hilborn's subcritical (100 W) design proposed a Sr–Be neutron source in the centre of a homogeneous light water core of uranium sulphate solution, subcritical by about −10 mk. The proposal included the feature suggested by Jarvis and Mills — the use of a substantial beryllium reflector to minimize the subcritical mass of ^{235}U, estimated to be $\approx$480 g. The neutron source would then amplify the fission flux from the subcritical core. Control was proposed to be self-regulating by varying the core temperature, making use of the negative temperature coefficient. This early conceptual idea of a low cost fission neutron source [6] was not pursued but led immediately to three years of further development followed by the design, construction and testing of the SLOWPOKE-1 prototype reactor by Hilborn's team, then at AECL, Chalk River Laboratories (CRL) [7] (Fig. 1). The first criticality of this 5 kW prototype was on 8 May 1970, with subsequent power transient tests reaching 186 kW (see Section 2.2.4). The basic feature of the first TRIGA Mark I reactors

— thermal flux $\approx 10^{13}$ cm$^{-2}\cdot$s^{-1} and thermal power of 250 kW, offering a high degree of safety from reactivity insertions — was also paramount in informing the SLOWPOKE-1 design.

A number of novel research reactor features were introduced by Hilborn, who was aware that the TRIGA reactor, although technically excellent [8], had high operating costs, driven by both the cost of fuel and the operational supervision requirements. The novel SLOWPOKE-1 general features focused on achieving (i) simplicity in both construction and low long term operating costs; (ii) a high degree of reactivity safety; and (iii) a high ratio of maximum thermal neutron flux to thermal power[2] using natural circulation cooling. Some of the main SLOWPOKE design and operation features were:

FIG. 1. Left: SLOWPOKE-1 prototype reactor installation into the Pool Test Reactor, Chalk River, May 1970 (courtesy of AECL and Canadian Nuclear Heritage Museum, Canada). Right: MNSR-IAE prototype reactor container, 1983 (courtesy of CIAE, China).

---

[2] SLOWPOKE-2 has among the highest thermal neutron flux to power ratio of any research reactor (subcritical and critical assemblies not considered for this effect).

— A low critical mass of 760 g ^{235}U providing a long core life of more than 20 years, enabled by the periodic addition of beryllium plates (shims) acting as a top reflector;
— No need for an engineered mechanical automatic shutdown system;
— No expensive and complex multidecade neutronics instrumentation, as compared with other low power research reactors;[3]
— No need for a leaktight containment building for prevention of release of fission products;
— A simple fuel design;
— Access to the core for operating staff and users strictly limited, and not required for normal operation and utilization;
— Minimal operating staff levels, training requirements and operating costs;
— A simple single rod control system using a single self-powered flux detector (SPFD);
— Allowing the SLOWPOKE-2 to be licensed for off-site remotely attended operation, with no operators in attendance at the reactor, with the reactor at full power for up to 24 hours for NAA and isotope production purposes (see Section 3.10);
— Possibility to perform irradiations outside the fuel core in the beryllium radial reflector high thermal neutron flux.

Almost all of these were also features of MNSRs, although the MNSR used an automatic shutdown system and multidecade neutronics instrumentation. It is noted that the lack of these two design features for SLOWPOKE is not consistent with current IAEA safety standards, namely IAEA Safety Standards Series No. SSR-3, Safety of Research Reactors [9]. However, aspects of research reactor safety design were not dealt with in IAEA safety standards until the 1980s; guidance on safe operation was the main focus before that time. The operational feature of allowing high power operation with remotely attended operation and with no licensed operator in attendance was also not foreseen or considered in the 1970s IAEA safety standards. When these two design features and the remotely attended operation feature were introduced in SLOWPOKE reactors in the early 1970s, they were recognized as being unique for research reactors at that time. Early SLOWPOKE licensing therefore addressed these three particular features with appropriate operational limits and conditions (OLCs). The OLCs and the early safety analysis reports (SARs) provided the basis for national regulatory acceptance in the context of the best practices at that time (see Section 3.10).

---

[3] The CRL prototype SLOWPOKE-1 was provided with conventional neutronics for test purposes. The SLOWPOKE-2 reactors all came without conventional multidecade neutronics instrumentation.

These original features remain to date and have been incorporated pragmatically into the licensing process.

The features of lack of core access and single control rod operation closely followed E. Teller's requirements for a safe reactor [2]. The focus on simplicity was perhaps the reason no patents were taken out on the SLOWPOKE design. Hilborn had implied the reactor design was too simple to patent [10].

The first research reactors in China had medium to high power levels, in the range 2–125 MW [11]. These reactors were located outside cities, and operating costs were found to be expensive, particularly for frequent users of NAA and producers of short lived isotopes. The capability of providing quick and frequent sample irradiations with a reactor that was simple and inexpensive to operate was thus found to be suitable for China at that time. Following the success of SLOWPOKE, in the late 1970s the China Institute of Atomic Energy (CIAE) began to develop the prototype MNSR-IAE, rated at 27 kW [12, 13].

As with SLOWPOKE reactors, the main applications for MNSRs were intended to be thermal NAA, production of medium and short lived isotopes, and education and training for nuclear engineers and technicians. The Institute of Geochemistry of the Chinese Academy of Sciences, the Institute of Geology of the Ministry of Geology, the Shandong Gold Geological and Mineral Exploration Company, the Daqing Oilfield Construction Design and Research Institute and the Institute of High Energy Physics of the Chinese Academy of Sciences showed great interest in carrying out NAA activities. A significant MNSR development project at the time was the zero power MNSR testing performed at the CIAE in Beijing in October 1981. In 1982, construction of the prototype MNSR-IAE reactor building and reactor support systems was completed [14], followed in 1983 by the commissioning of the control system, thermal measurement analysis and dose monitoring in the reactor building and the surrounding environment [15]. The MNSR-IAE reached first criticality on 10 March 1984. Prototype commissioning testing with a maximum power transient of 3.6 mk was performed on 5 December 1984 [14, 15]. This produced a peak power of 76 kW, consistent with SLOWPOKE-2 power transient tests [16]. Soon after commissioning, the MNSR-IAE reactor was operated by researchers without an on duty operator [11]. In 1986, Hilborn and design colleagues met staff from Beijing's Institute of Nuclear Engineering at AECL CRL to discuss mutual interest in SLOWPOKEs and MNSRs with regard to potentially upgrading the reactor designs for commercial district heating purposes.

Four more MNSRs were subsequently manufactured by CIAE for use in China from the mid-1980s on [17]. These were the MNSR-SZ at Shenzhen University, which reached criticality in 1988; the MNSR-SD in Shandong in 1989; the MNSR-SH in Shanghai in 1991; and the IHNI-1 in Fangshan in 2009. Five more were built outside China: in Pakistan in 1989, the Islamic Republic of

Iran in 1994, Ghana in 1994, the Syrian Arab Republic in 1996 and Nigeria in 2004. One MNSR is under construction in Thailand. The MNSRs built after the prototype MNSR-IAE are referred to in this publication as commercial MNSRs.

Tables 1 and 2 provide an overview of the SLOWPOKEs and MNSRs. Table 1 lists the operational and under construction SLOWPOKEs and MNSRs. Table 2 lists those shutdown.

TABLE 1. OPERATIONAL AND UNDER CONSTRUCTION SLOWPOKE AND MNSR REACTORS

IAEA code	Country	Facility name and location	Nominal power (kW)	First criticality
CA0009	Canada	SLOWPOKE-2, Polytechnique Montréal, Quebec	20	1 May 1976
CN0006	China	MNSR-IAE Prototype, Department of Reactor Engineering & Technology, CIAE, Fangshan District, Beijing	27	10 Mar. 1984
JM0001	Jamaica	SLOWPOKE-2, International Centre for Environmental and Nuclear Sciences, University of the West Indies, Mona Campus	20	13 Mar. 1984
CA0014	Canada	SLOWPOKE-2, Royal Military College, Kingston, Ontario	20	6 Sep. 1985
CN0013	China	MNSR-SZ, Joint Institute of Applied Nuclear Technology, Shenzhen University, Shenzhen, Guangdong	30	1 Nov. 1988
PK0002	Pakistan	PARR-2, MNSR, Pakistan Institute of Nuclear Science & Technology, Islamabad	30	2 Nov. 1989
IR0005	Iran, Islamic Republic of	ENTC MNSR, Isfahan Nuclear Technology Centre, Isfahan	30	1 Mar. 1994

TABLE 1. OPERATIONAL AND UNDER CONSTRUCTION SLOWPOKE AND MNSR REACTORS (cont.)

IAEA code	Country	Facility name and location	Nominal power (kW)	First criticality
GH0001	Ghana	GHARR-1, MNSR, Atomic Energy Commission, National Nuclear Research Institute, Accra	34	17 Dec. 1994
SY0001	Syrian Arab Republic	SRR-1, Syrian MNSR, Atomic Energy Commission, Damascus	30	4 Mar. 1996
NG0001	Nigeria	NIRR-1, MNSR, Ahmadu Bello University, Centre for Energy Research and Training, Zaria	34	3 Feb. 2004
CH0019	China	IHNI-1, MNSR, Beijing Capture Technology Co., Fangshan District, Beijing	30	7 Dec. 2009
TH003	Thailand	SUT MNSR, Suranaree University of Technology, Nakhon Ratchasima	45	Under construction

Note: The IAEA code is taken from the IAEA's Research Reactor Database (RRDB).

TABLE 2. SHUTDOWN SLOWPOKE AND MNSR REACTORS

IAEA code	Country	Facility name and location	Nominal power (kW)	First criticality	Permanent shutdown/ decommissioned status
—[a]	Canada	SLOWPOKE-1, AECL, Chalk River Laboratories, Chalk River, Ontario	Transient tested to 186	8 May 1970	This was the original SLOWPOKE-1 reactor. Relocated to the University of Toronto in April 1971.
—	Canada	SLOWPOKE-1, University of Toronto, Ontario[b]	5–20	4 Jun. 1971	A new HEU core was used. For decommissioning see footnote b.

TABLE 2. SHUTDOWN SLOWPOKE AND MNSR REACTORS (cont.)

IAEA code	Country	Facility name and location	Nominal power (kW)	First criticality	Permanent shutdown/ decommissioned status
CA0008	Canada	SLOWPOKE-2, University of Toronto, Ontario[c]	20	10 Mar. 1976	Shutdown: 31 Dec. 1998. Decommissioning Licence: 5 Oct. 2000. Licence to abandon: 8 Feb. 2001.
—	Canada	SLOWPOKE-2, AECL Commercial Products, Tunney's Pasture, Ottawa, Ontario	20	1 May 1971	Reactor was relocated to Kanata in 1984.[d] Shutdown: 11 Jan. 1984.
CA0010	Canada	SLOWPOKE-2, Dalhousie University, Halifax, Nova Scotia	20	8 Jul. 1976	Decommissioning authorized: 20 Jan. 2011. Licence to abandon: 31 Aug. 2011.
CA0011	Canada	SLOWPOKE-2, University of Alberta, Edmonton, Alberta	20	22 Apr. 1977	Decommissioning authorized: 22 Sep. 2017. Licence to abandon: 25 May 2018.
CA0012	Canada	SLOWPOKE-2, Saskatchewan Research Council, Saskatoon, Saskatchewan	20	13 Mar. 1981	Decommissioning authorized: 6 Dec. 2019. Licence to abandon: 1 Oct. 2021.
CA0013	Canada	SLOWPOKE-2, Kanata Isotope Production Facility, MDS Nordion, Kanata, Ontario (iv)	20	6 Jun. 1984	Shutdown: 1 Apr. 1989.

TABLE 2. SHUTDOWN SLOWPOKE AND MNSR REACTORS (cont.)

IAEA code	Country	Facility name and location	Nominal power (kW)	First criticality	Permanent shutdown/decommissioned status
CA0016	Canada	SLOWPOKE Heating Demonstration Reactor (SDR), AECL, Whiteshell Laboratories, Manitoba	2000	15 Jul. 1987	Shutdown: 12 Sep. 1989. Decommissioning completed: Apr. 2016.
CN0014	China	MNSR-SD, Shandong Institute of Geological Sciences, Jinan, Shandong	33	1 May 1989	Shutdown: 2008. Decommissioning authorized: Dec. 2010 Building free release authorized: Feb 2012.
CN0015	China	MNSR-SH, Shanghai Institute for Measurement and Testing Technology, Shanghai	30	15 Dec. 1991	Decommissioned: 2006–2008.

[a] Not included in the RRDB.

[b] The transfer of the reactor container was completed using a wooden shipping container. On 15 Jan. 1976 the SLOWPOKE-1 reactor container and the fuel core were removed and transported separately to AECL CRL in an operation taking only two days.

[c] This new SLOWPOKE-2 reactor was installed on 22 Feb. 1976 and a new fuel core installed on 10 Mar. 1976.

[d] This SLOWPOKE-2 reactor was relocated from Tunney's Pasture in 1984, with a new HEU fuel core being installed.

2.2. PROTOTYPE STUDIES FOR SLOWPOKE AND MNSR DEVELOPMENT

This section includes schematics and photographs of a number of SLOWPOKEs and MNSRs and their components.

2.2.1. Basis for fuel

This section provides an overview of the basis for fuel for SLOWPOKEs and MNSRs for both HEU and LEU fuel.

2.2.1.1. High enriched uranium

The original fuel for the SLOWPOKE reactors, except for the Royal Military College (RMC) reactor, was HEU with $\approx$ 0.82 kg of 93% ^{235}U and 28 wt% uranium–aluminium (U–Al) alloy. A description of the history of this choice of HEU enrichment is provided in Ref. [17]. The fuel design was based on the licensed design for the Canadian National Research Experimental reactor (usually known as NRX reactor), the National Research Universal reactor (usually known as NRU reactor) and other reactors [18, 19]. The early fuel for the NRX and NRU reactors had many years of satisfactory operational experience with low temperature water cooling and to much higher burnups and power ratings than needed for the SLOWPOKE design. As a result, there was no requirement for an irradiation testing programme for the HEU SLOWPOKE fuel.

All MNSRs in Tables 1 and 2 were also originally fuelled by U–Al alloy containing 90% enriched ^{235}U, except for the SUT MNSR under construction, which will use LEU. The HEU fuel and its aluminium alloy 303-1 cladding demonstrated satisfactory irradiation performance with no corrosion problems. The MNSR HEU fuel element design was similar, but not identical, to the SLOWPOKE HEU fuel (see Section 3.2.1).

2.2.1.2. Low enriched uranium

Since the late 1970s, there have been international efforts to reduce the use of HEU fuel in research reactors in order to reduce the global proliferation risk. As early as 1977, a development programme for SLOWPOKE LEU fuel was suggested by Hilborn, and in 1983 UO$_2$ LEU SLOWPOKE fuel was first manufactured [17, 20–22]. UO$_2$ has a high melting point, good chemical stability, good compatibility with Zircaloy cladding and excellent irradiation stability.

LEU was used for the first fuel charge of the RMC SLOWPOKE in 1985, as well as for the subsequent 1997 LEU conversion of the Polytechnique Montréal SLOWPOKE and the 2015 LEU conversion of the Jamaica SLOWPOKE. All three currently operational SLOWPOKE reactors are now using identical LEU fuel. Feasibility studies for the MNSR IHNI-1 LEU fuel were started in 2004 and LEU fuel fabrication was completed in 2006. MNSR conversions were started under the Reduced Enrichment for Research and Test Reactors programme in 2005 [23] with a LEU fuel element design similar to the MNSR IHNI-1 LEU.

The LEU fuel element design for both SLOWPOKEs and MNSRs is based upon the well established and reliable Canada Deuterium Uranium (CANDU) UO_2 ceramic fuel element design, but with a smaller outside diameter and different types of end cap (see Section 3.2). As with the HEU, no fuel test irradiation programme was needed for the LEU because of the extensive operating experience with CANDU fuel. The fuel element cladding and the fuel cage assembly (the core fuel stage comprises lower and upper flanges and tie rods for supporting the fuel elements and rests on the lower beryllium reflector) are both made of Zircaloy-4 alloy, which has a small neutron cross-section and good water corrosion resistance. Helium gas filling between the cladding and fuel is used to improve thermal conductivity between the cladding and the fuel meat.

2.2.2. Basis for moderator and coolant

The fuel cages for both HEU and LEU fuel core assemblies are immersed in light water inside the reactor container. This water acts as a neutron moderator, outer reflector and heat transport coolant. The moderator and coolant water extend vertically about 4.4 m above the top of the fuel cage and provide adequate radiation shielding for personnel access at the top of reactor at the floor level of the reactor room (Fig. 2; see also Ref. [24]). The reactor container water is entirely separate from the reactor pool water (see Fig. 2) and uses a dedicated purification system to maintain adequate pH, conductivity and activity levels.

The low reactor power allows the moderator and coolant water to operate under a natural convection unpressurized condition with no forced cooling required for core heat removal. Heat from fuel elements is transferred to the cooling water in the core by means of thermal conduction and natural convection heat transfer. The water is heated within the volume of the fuel cage and flows upwards by natural convection. Then, after flowing from the outlet of the reactor core, the cooling water gradually rises to the upper space of the reactor container. At the same time, some heat is transferred to the pool water by conduction through the reactor container. Cooler water along the inner wall of the reactor container then flows down along the inner wall and returns to the bottom of the core inlet (Fig. 3).

2.2.3. Basis for reflector

The SLOWPOKE fuel cage is inserted into a 10.2 cm thick, 22.8 cm high annular beryllium reflector. The fuel cage sits on top of a lower 10.2 cm thick beryllium cylindrical disc, 32.2 cm in diameter (Figs 3 and 4). The MNSR fuel cages are similarly inserted within 10.2 cm thick and 24.8 cm high annular beryllium reflectors. A top reflector, consisting of semicircular beryllium plates

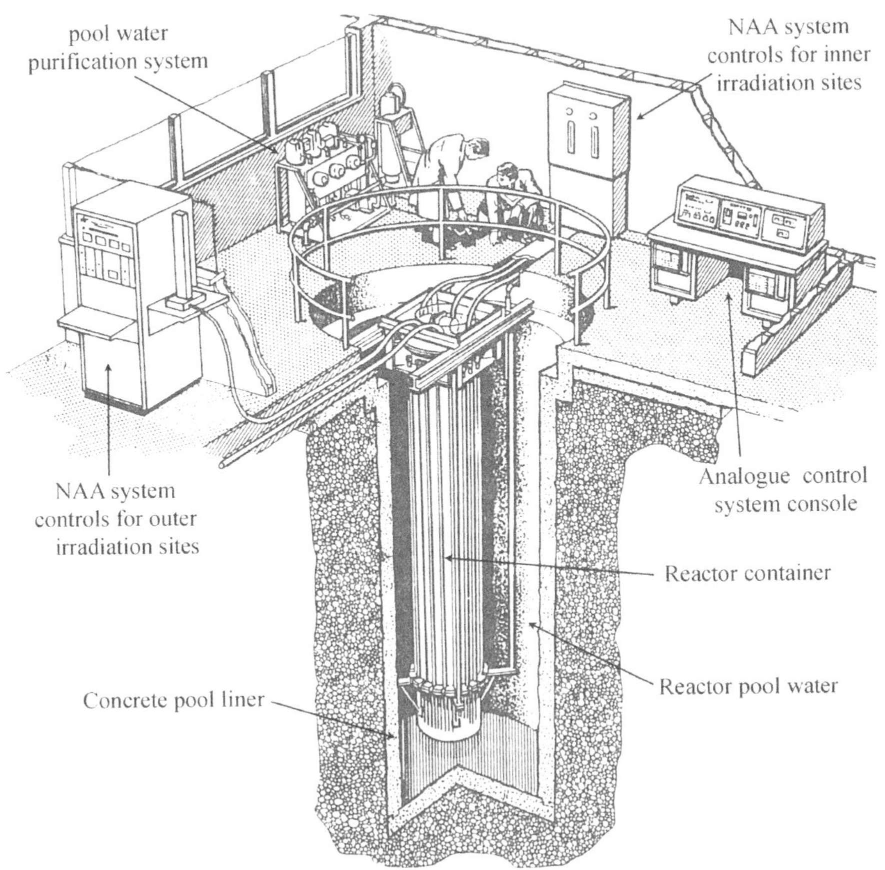

FIG. 2. *SLOWPOKE reactor installation at Tunney's Pasture [24] (courtesy of AECL and Canadian Nuclear Heritage Museum, Canada).*

called shims, each a few millimetres thick, sits on an aluminium shim tray above the fuel cage. The system excess reactivity decreases in the long term due to ^{235}U burnup and long lived fission product poison buildup, mainly from ^{149}Sm. The equilibrium concentration of ^{149}Sm builds up over long term operation, is independent of neutron flux and power level, and is worth about −5.5 mk. This long term reactivity decrease is compensated for by periodic additions of shims to the top beryllium reflector. The addition of these shims reduces neutron leakage above the fuel cage as the top beryllium reflector overall thickness is increased. Experience has demonstrated that such a shim adjustment is usually required every 3–4 years for high utilization operating cycles, sometimes less frequently, depending on the operational usage history [25] (see Section 3.6.1.7).

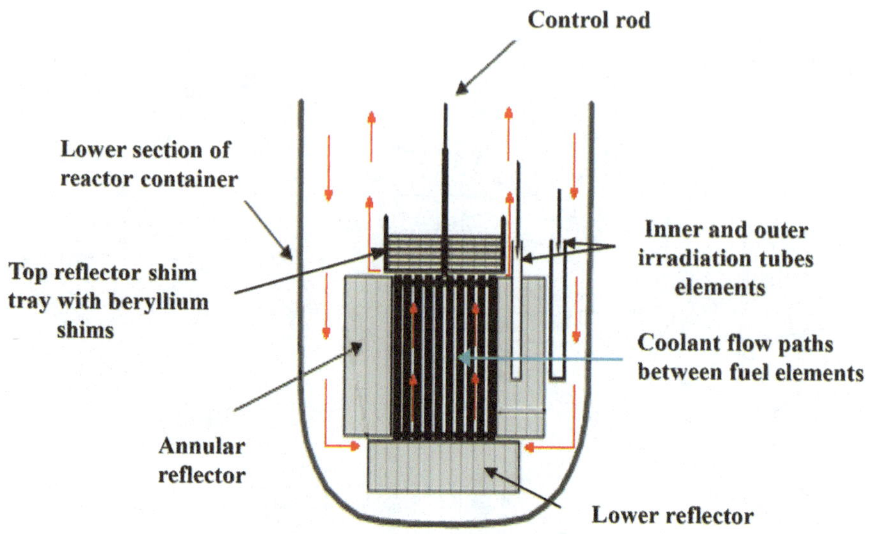

FIG. 3. Natural circulation coolant flow paths (red) in lower reactor container section (courtesy of D.J. Winfield, Canada).

As fuel burnup increases and the thickness of the beryllium top reflector is correspondingly increased, the incremental reactivity worth per centimetre of additional beryllium shims decreases. A typical addition of one 3.2 mm thick semicircular beryllium shim is worth about 0.7 mk on a fresh core. The semicircular beryllium plates are arranged on the shim tray most effectively when both halves of the tray shims are level. The maximum shim thickness is about 10 cm and is typically worth up to a total of about 18 mk for either the HEU or the LEU cores. Similarly, the MNSR total worth of the upper beryllium reflector is about 16 mk. At two SLOWPOKE reactors, Dalhousie University in 1987 and the University of Toronto in 1988 (see the Annex to this publication), core lifetimes were extended an extra 10 years by adding an additional beryllium annulus reflector around the core. This additional beryllium was known as a 'big shim' (also referred to as a 'doughnut reflector'). It was manufactured to fit around the existing beryllium annulus reflector, with holes to allow for the presence of the outer irradiation tubes.

2.2.4. Basis for reactor control philosophy

From its inception, the SLOWPOKE reactor control and safety philosophy has been to tailor the engineering design and operating procedures with various inherent safety characteristics to give a high degree of reactivity safety for

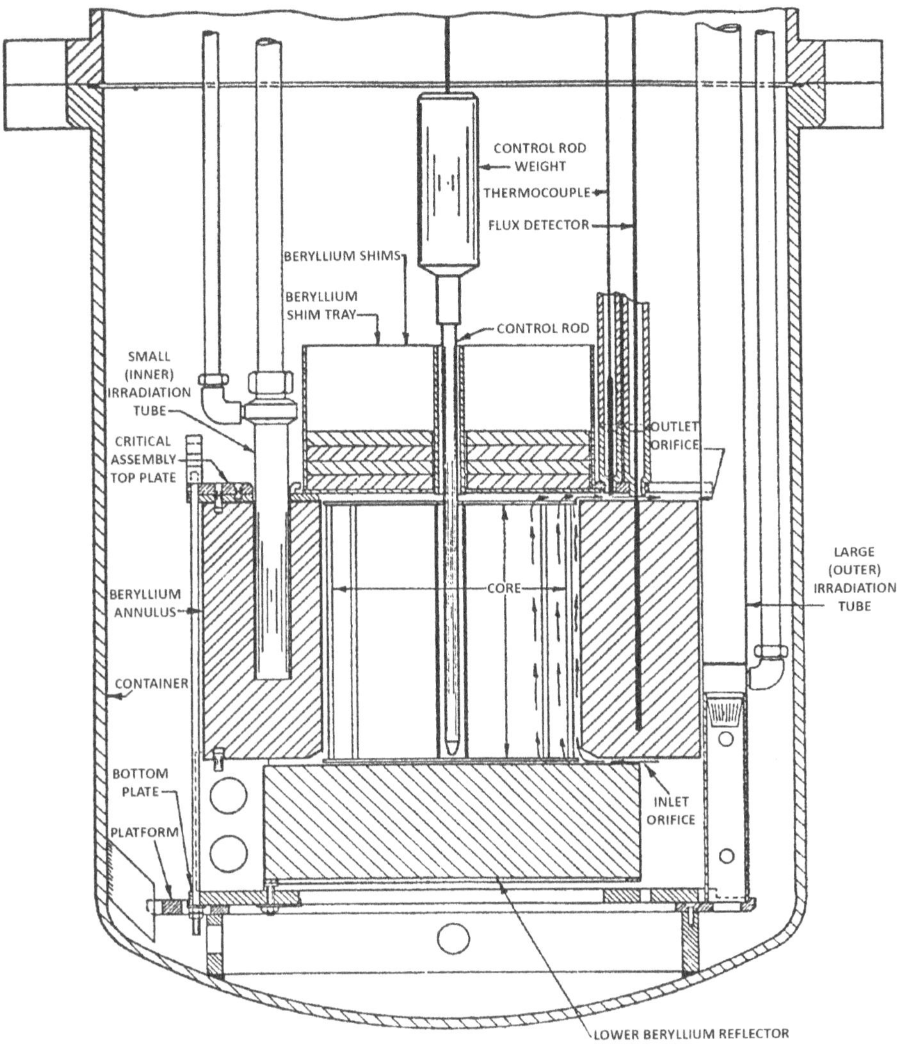

FIG. 4. Vertical cross-section through the SLOWPOKE-2 reactor container [19] (courtesy of AECL and Canadian Nuclear Heritage Museum, Canada).

design basis reactivity insertion accidents. The conventional, redundant channel automatic electromechanical shutdown system used for all other research reactors and all power reactors was not used. The most important contributing safety features are the negative power coefficient and void coefficients, available from the combined effect of the fuel and moderator and coolant temperature coefficients, leading to self-limiting power excursions from large reactivity insertions. Section 3.6 provides information on the power transient testing of the

15

CRL SLOWPOKE-1 prototype reactor. Beyond design basis reactivity insertions up to 6.48 mk were performed at that reactor, leading to a maximum peak self-limiting power of 186 kW after a few minutes. This peak power provides an adequate safety margin for HEU, being well below the predicted critical heat flux (CHF) limit of the natural circulation core and also well below centre line melting of the fuel (see Section 3.6.2.1 for details).

The cornerstone of the safety philosophy is to carefully limit the accessible maximum excess reactivity of SLOWPOKE reactors to 4 mk (well below the maximum reactivity insertion transient test of the original prototype) under all conceivable conditions. This value was updated to 4.3 mk by the Canadian Nuclear Safety Commission (CNSC) in the relicensing of the RMC SLOWPOKE-2 in 2024. Only one central control rod was then needed for reactivity control. SLOWPOKE-2 reactors do not use any automatic shutdown system, although the CRL prototype SLOWPOKE-1 was provided with conventional automatic shutdown neutronics for test purposes. The SLOWPOKE-1 reactors used both a manual and an analogue autocontrol system [26]. Manual control used a curved cadmium absorber plate, clad in aluminium, located close to the outer surface of the radial beryllium reflector, connected to a manually operated, motor driven lead screw. The SLOWPOKE-1 analogue autocontrol system used a central control rod design that was later installed in all SLOWPOKE-2 reactors. The purpose of the manual control plate was to shut down the reactor if the central control rod failed in the fully up position.

The MNSR control system has some differences compared with that of the SLOWPOKE. The MNSRs provide an automatic shutdown on two parameters: high reactor power and high temperature difference between the fuel cage coolant inlet and outlet orifices (see Fig. 3). The MNSR control rod is of a similar design but is not identical to the SLOWPOKE-2 control rod. The main difference is that it also operates as an automatic shutdown system rod using an electromagnetic clutch holding up the control rod cable drive and dropping the control rod on loss of clutch power. MNSRs also have two to four reactivity adjuster rods that are used to adjust the initial core excess reactivity to the required level (see Section 3.9 and Ref. [27]). The reactivity adjuster rods are not connected to the control rod drive system or to the control system, but rather are fixed in place.

2.2.5. Historical developments from the original concept

Since the original SLOWPOKE-1 and MNSRs were built, a number of developments and modifications have been pursued that have advanced the original design concept. There have also been proposals that have not been pursued or implemented. Both developments (demonstrations) and proposals are summarized in Sections 2.2.5.1–2.2.5.12, in chronological order. Minor design changes and improvements to various systems are not included. Control system changes to digital computer control (DCC) are discussed in Sections 3.5.1 and 3.5.2, and fuel conversions are discussed in Section 5.2.

2.2.5.1. *Direct SLOWPOKE-2 reactor container cooling, 1981*

A modification was made to the University of Toronto SLOWPOKE-2 reactor to demonstrate an increase in the available continuous operational time by limiting the coolant temperature rise during operation at power [28]. The reactivity increase produced by limiting the coolant temperature showed that such cooling could be used as a routine procedure without compromising reactivity safety. An auxiliary cooling system was installed that injected chilled water from a refrigeration unit directly into the reactor container via nozzles directed towards the coolant inlet orifice or via the discharge line of the reactor container purification system. An increase in the continuous operational time at full power, up to 8 hours and more, was demonstrated.

2.2.5.2. *SLOWPOKE reactor for hybrid submarine design, 1985*

In the mid-1980s, the Canadian Government considered using a SLOWPOKE reactor to provide continuous recharging of the batteries of the British-designed Oberon class submarines used by the Canadian navy, allowing for longer submerged operational periods [29]. A nuclear battery–electric hybrid was suggested to be a potentially attractive alternative to the dominant nuclear turbomechanical drive. The proposal was not realized.

2.2.5.3. *SLOWPOKE demonstration reactor, 1987*

Using some basic design concepts of the SLOWPOKE-2, a 2 MW(th) reactor, the SLOWPOKE Heating Demonstration Reactor (SDR), was built at the AECL Whiteshell Laboratories in the period 1985–1987, intended to demonstrate the technical, economic and safety criteria needed to provide local nuclear district heating as an alternative to the use of oil and diesel fuel in remote locations (see Annex). In addition, the intent was to provide prototype testing of concepts and systems to support a higher power version, the SLOWPOKE Energy System SES-10 programme (see Section 2.2.5.5 and Annex). The SDR reactor used 4.9% ^{235}U LEU with 4 UO_2 fuel bundles. The reactor was successfully commissioned up to 1 MW (50% nominal full power) by May 1989 but was shut down shortly afterwards, and was eventually decommissioned when the economic and political climate for nuclear district heating in Canada became unfavourable.

2.2.5.4. *MNSR-SD addition of two beam tubes, 1989*

Like the SLOWPOKE design, the original MNSR design did not include beam tubes. The MNSR-SD installed two dry vertical neutron beam tubes, 5 cm in diameter, outside the reactor container, which were used for neutron detector testing and calibration (see Annex).

2.2.5.5. *SLOWPOKE energy system-10, 1990*

The AECL initiated a further scaleup of the 2 MW(th) SDR SLOWPOKE to a 10 MW(th) version called the SLOWPOKE Energy System SES-10 (see Annex). This version was mainly intended for district heating in an urban environment. Significant design work was performed. The design, based on SDR experience, did not advance to the regulatory approval stage and was never built. The project was cancelled for the same reasons as the SDR (see Annex).

2.2.5.6. *MNSR IHNI-1 boron neutron capture therapy facility, 2009*

The first basic MNSR design modified to be suitable for hospital siting by providing neutron beams for boron neutron capture therapy (BNCT) for patient cancer treatment was the IHNI-1 reactor [30, 31].[4] The Beijing Capture Technology Company operates the facility, which was constructed during 2001–2010. Fuel loading approval was received from China's National Nuclear Safety Administration (NNSA) in October 2009. Full power was achieved in January 2010. The first patient was treated using the IHNI-1 in September 2014; the reactor had more than 150 operational periods between 2010 and 2020 [32]. The IHNI-1 installed three neutron beam tubes (two horizontal and one vertical) and was the first MNSR fuelled with LEU for the first fuel charge.

2.2.5.7. *RMC SLOWPOKE addition of a radiography beam tube, 2013*

A significant modification, made by the SLOWPOKE vendor for the RMC, was the installation of a thermal column, replacing an irradiation site inside the reactor container and allowing installation of a neutron beam tube to be used for a neutron radioscopy system using digital imaging [33]. The CNSC approved reactor operation of this neutron beam tube with the reactor at full power in 2013.

---

[4] A proposal in 2000 to build a SLOWPOKE BNCT facility was not followed up (see https://www.nuceng.ca/univcomm/studconf2000/paper1-2-mccall.pdf).

2.2.5.8. Homogeneous core SLOWPOKE for ^{99}Mo production, 2014

A homogeneous SLOWPOKE reactor was proposed [34] to replace the heterogeneous SLOWPOKE-2 reactors when they reached end-of-core life, mainly for NAA and also to produce radioisotopes such as ^{99}Mo/^{99m}Tc for medical applications. The homogeneous SLOWPOKE reactor was designed to use a liquid fuel core that would enable these isotopes to be extracted. A homogeneous fuel mixture of dilute aqueous solution of uranyl sulphate with about 1 kg of 20% ^{235}U using a Zircaloy-2 reactor container was proposed. To date the proposal has not been followed up.

2.2.5.9. ENTC MNSR addition of radiography beam tubes, 2017–2018

Similar to the installations of other beam tubes (see Sections 2.2.5.4 and 2.2.5.6), the ENTC MNSR also installed a beam tube in the pool, to be used for neutron radiography (see Annex). A vertical aluminium tube, 5 cm in diameter, was first installed in the ENTC MNSR for prototype validation measurements. A larger 25 cm diameter tube was then installed for practical radiography usage.

2.2.5.10. Organic-cooled SLOWPOKE reactor, 2018

There was a 2018 proposal to develop a reactor based on the SLOWPOKE-2 research reactor, but cooled and moderated by an organic fluid. The intended use was heating applications in northern Canada [35]. The conceptual design considered using the organic coolant HT-40 instead of light water, allowing a thermal power of 1 MW. The dimensions of this organic SLOWPOKE-2 were to be enlarged, allowing for a larger mass of uranium and providing a larger surface area for more efficient heat transfer. Notwithstanding these proposed modifications, the design retained the simplicity of the SLOWPOKE-2 design. The proposal was not followed up.

2.2.5.11. ENTC MNSR addition of a prompt gamma beam tube, 2022–2023

In addition to the two beam tubes for neutron radiography (see Section 2.2.5.9), the ENTC MNSR installed a third beam tube close to the reactor core for prompt gamma ray analysis (PGA)[5] (see Annex and Ref. [36]). The PGA beam tube is a vertical aluminium beam tube with a diameter of 3.9

---

[5] 'Prompt gamma ray analysis' is considered the current term in this publication, whereas 'prompt gamma activation analysis' and 'prompt gamma neutron activation analysis' are considered deprecated.

cm and a height of 530 cm. It is installed inside the reactor tank and behind the annular beryllium reflector in one of the four positions initially considered for the installation of reactivity regulating rods. Only one of these positions was occupied by the reactivity regulating rod, and the other three positions were available. To minimize background radiation, the neutron radiography and PGA beam tubes are not to be used simultaneously.

2.2.5.12. MNSR boron neutron capture therapy facility, from 2022

The Suranaree University of Technology research reactor (SUT MNSR), Thailand, has been under development since 2022. It is an LEU fuelled MNSR design, upgraded to 45 kW (see Annex). The main purpose of the reactor is for BNCT. In addition to having the highest MNSR nominal power, it is designed with five neutron beams. Four of these beams are horizontal, for neutron radiography, NAA and PGA research. The other beam is vertical, below the core, for BNCT.

3. TECHNICAL CHARACTERISTICS

This section describes the technical characteristics of SLOWPOKEs and MNSRs. The dimensions it gives for both are an adequately accurate representation of each of the two groups. The exceptions are the prototype SLOWPOKEs and MNSRs and the SUT MNSR, which is not yet built and is expected to have several design evolutions. Minor differences, such as the dimensional and structural differences between the main components of the MNSR and SLOWPOKE-2 reactor containers (see Section 3.1.3), are not discussed, as they are considered inconsequential within the scope of this publication.

3.1. REACTOR DESIGNS

This section provides information on different aspects of SLOWPOKE and MNSR original designs, including the reactor building, the reactor pool, the reactor container, the reactor critical assembly and associated support systems. It also provides information on modifications to the original design. Figures 2–4 illustrate the location and schematic of each component (see Fig. 6 for fuel cages).

3.1.1. Reactor building

The original generic SLOWPOKE-2 description and safety analysis [19] did not specify any design requirements for a reactor building. Facility building designs were provided by each facility to supplement the vendor-supplied generic SAR. The high degree of reactivity safety limits the potential for reactivity initiated power transients. As a result, there were no facility-specific design requirements for a dedicated confinement or containment building,[6] typical of most other research reactors of higher power. All SLOWPOKE-2 reactors (see Tables 1 and 2) had the option of being accommodated within an existing building structure.

Similarly to the SLOWPOKE reactor, the MNSR also had a generic SAR [37]. For the same reasons as the SLOWPOKE reactors, it had no specific design requirements or guidance for a reactor building. All MNSRs are located in national nuclear research institutes, except for MNSR-SH, MNSR-SD, MNSR-SZ and IHNI-1, all of which are in China, and the Thai SUT MNSR (see Tables 1 and 2), and all have been accommodated into existing building structures.

3.1.2. Reactor pool

An adequately sized pool is needed to accommodate the reactor container and to provide radiation protection from core radiation, cooling of the reactor container coolant via heat transfer through the immersed container and underwater space for active component removal operations (see Fig. 2). The pool water has no contact with the fuel or coolant (see Fig. 4), so the risk associated with a leak of active reactor container water into the pool water and then out of the pool is very small. There are therefore no restrictive requirements for active liquid releases from the pool, but a leaktight pool design is of course still required. The various facility owners provided their pool designs and, as with the building support services, the owners provided the pool design description for their SARs. Circular pools with concrete peripheries and bases have been used for most SLOWPOKE-2 reactors. The RMC SLOWPOKE concrete pool also included a stainless steel liner. The University of Alberta pool was rectangular and made of reinforced concrete lined with ceramic tile. The MNSR-IAE,

---

[6] Requirement 43 of IAEA Safety Standards Series No. SSR-3 [9] states: **"Means of confinement shall be provided for a research reactor to ensure, or to contribute to, the fulfilment of the following safety functions: (i) confinement of radioactive substances in operational states and in accident conditions; (ii) protection of the reactor against natural external events and human induced events; and (iii) radiation shielding in operational states and in accident conditions."**

MNSR-SZ and MNSR-SH used circular concrete pools lined with ceramic tile. Other MNSR facilities, except for PARR-2, all used circular stainless steel pool liners. PARR-2 used a concrete rectangular pool, larger than all other MNSR and SLOWPOKE reactors, with a stainless steel liner and also a stainless steel base (see Annex). The original SLOWPOKE-1 reactor container was located in an existing concrete pool constructed for an earlier reactor. All currently operating reactors are subgrade, so radial biological shielding from the core is unnecessary. Typical pool dimensions have been $\approx$ 6.5 m in depth and $\approx$ 2.7 m in diameter, with pool water volumes of $\approx$ 30 m^3 for circular pools.

In the 1970s, SLOWPOKE-2 reactors in one storey buildings did not use concrete pool covers. The original design implied that pool covers were optional for the facility owner. Loss of pool water below the reactor container level would remove the vertical pool water shielding. Loss of water from subgrade concrete pools was considered a beyond design basis accident. If a room above the reactor in a multistorey building was occupied, higher than normal radiation fields would be possible if the pool water level was reduced for reasons other than a small leak through the pool walls. In multistorey buildings, such as the Polytechnique Montréal, which has offices above the reactor, 75 cm thick concrete radiation shielding covers over the pool and the reactor container were provided. Over time, the covers were then found to be necessary to keep radiation fields above the HEU reactor container at acceptable levels. This was due to slowly increasing radiation levels in the reactor container water and in the reactor container's air filled gas space after the reactors had operated at high power for a few hours, as some of the HEU fuel had developed fission product releases over time [38, 39]. The source of these releases was thought to be uranium-bearing material in some of the element end welds from the time of fuel fabrication. A band of exposed fuel was observed in a metallurgical examination of an unirradiated archived HEU element. None of the HEU SLOWPOKE reactors experienced any fission product activity in the pool water. For the JM0001 HEU SLOWPOKE reactors, the radiation fields above the pool were historically small enough that concrete covers were not required. The much lower activity levels in the container water of all current SLOWPOKE LEU reactors are likely limited to tramp uranium contamination on the Zircaloy element cladding.

None of the MNSR facilities have required concrete radiation shielding covers above the pool and the reactor container. As with LEU fuelled SLOWPOKE reactors, activity levels in the reactor container water of all commercial MNSRs have been limited to those resulting from tramp uranium contamination on the fuel cladding. The prototype MNSR-IAE reported some HEU fission product releases into the reactor container water (see Section 6.1), but no pool covers were found to be necessary. To prevent objects from falling into the pool and to minimize evaporation of pool water, polymethyl methacrylate pool cover plates are provided in MNSRs. These cover plates are supported with aluminium or steel frames. A sealing gasket is used between the supporting frame and the polymethyl methacrylate cover plate. The cover plate is designed for easy removal.

3.1.2.1. Pool water cooling system

The pool water has a large thermal capacity. If all the heat from reactor operation at 20 kW were absorbed by the pool water, the temperature rise — with no cooling — would be $\approx 0.6°C/h$ for facilities with the typical circular pool size of ≈ 3 m. The pool water temperature rise for 8 h of reactor operation at 10 kW (80 kWh) is calculated to be about 2.6°C.[7] To prevent the pool temperature from increasing to an undesirable level over a period of several days, the pool water for SLOWPOKEs is provided with a cooling system. All MNSRs also use a pool water cooler, except for IHNI-1 in China.

The pool water cooling system is a simple closed loop system comprising a heat exchanger immersed in the pool, isolating valves and a flow monitor. The type of heat exchanger varies according to the facility. Typically, building service water can supply an adequate flow rate of ≈ 30 L/min to maintain the pool water close to the ambient reactor room temperature. A refrigeration cooling unit can be used if pool temperatures below the ambient temperature are needed. There is a continuous but very slow loss of pool water due to evaporative losses into the reactor room. Typical maximum pool water make-up rates documented in facility annual reports, which are produced by all facilities,[8] have been ≈ 6 L/d over 30 years of operation for SLOWPOKE reactors and similar values are reported for MNSRs. Facilities in warm climate locations would have higher make-up rates. The pool level is maintained by automatic addition of service water to the pool purification system via a simple pool water level control system.

3.1.2.2. Pool water purification system

The pool water is continuously purified to minimize the corrosion of the outside of the reactor container by maintaining the required conductivity and pH specifications (see Section 3.8) [9]. Pool water purification systems vary between facilities but typically operate on a continuous basis of ≈ 10–20 L/min, using

---

[7] PARR-2, with a large pool volume of 126 m^3 (see Annex), would have a smaller temperature rate increase.

[8] Annual reports are currently provided to the various national regulators for all SLOWPOKE-2 reactors and MNSRs. Some Member States' annual reports are available publicly, some upon request and others are not available to the public.

ion exchange resins and, possibly, filters. The suction end of the inlet line is positioned ≈ 10–20 cm below the pool water surface. This suction level is chosen as a compromise to maximize good circulation mixing of the pool water without increasing potential radiation fields to an unacceptable level on top of the pool in the event of a purification system leak outside the pool, which could lower the pool water level. The discharge end of the purification system outlet line leads to the bottom of the pool.

3.1.3. Reactor container

The reactor assembly, comprising the core, reflectors, control rod, instrument probes and irradiation tubes, is housed inside a reactor container (see Figs 2 and 4). The reactor container is bolted closed with a top plate (Fig. 5) after the maximum excess reactivity has been established during original commissioning (see Sections 2.2.4 and 3.9). The container is periodically reopened for the addition of beryllium top reflector plates (shims) to maintain excess reactivity, or for the maintenance of internal components. There are some minor dimensional and structural differences between the MNSR and SLOWPOKE-2 reactor container main components; these are not discussed here, being inconsequential within the scope of this publication.

The water surface inside the reactor container is maintained at approximately the same level as the pool water. A top plate closes the top section of the reactor container, which leaves a small enclosed gas space (i.e. the headspace) between the top plate and the reactor container water. The reactor container thus provides a physical barrier between the reactor container water and the pool water. The reactor container is designed to be leaktight so that the container water will not mix with the pool water. To date, no leaks into pool water from the reactor container have been detected in any of the SLOWPOKE reactors or MNSRs. Information on historic radionuclide concentrations in the reactor container water and pool water can be found in annual reports.

The SLOWPOKE aluminium reactor container (see Fig. 4) has a 61 cm outside diameter and 1 cm thickness, and is constructed in two parts, with a lower section (83 cm long) and an upper section (4.4 m long); the fuel cage assembly is located within the lower section. Commercial SLOWPOKE and MNSR container dimensions are similar. In commercial MNSRs, the upper and lower sections are 4.7 m and 0.9 m long, respectively. The upper section is essentially an extension of the lower section, providing the depth of water necessary for effective radiation shielding above the pool from the fuel core. An aluminium seal ring is used at flanged joints between the two container sections to provide a watertight seal.

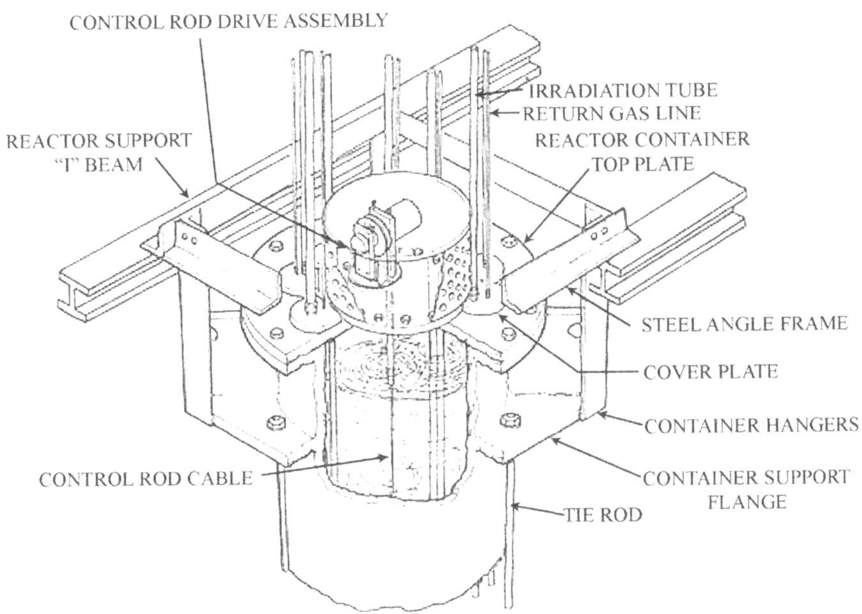

FIG. 5. SLOWPOKE reactor container mounting assembly showing container top plate and support beams; MNSRs are similar in their design (courtesy of AECL and Canadian Nuclear Heritage Museum).

The split, two section reactor container was designed to enable a spent core to be removed from the reactor at the end of its useful life, with adequate radiation shielding from the pool water still being in place. For all commercial SLOWPOKE reactors and MNSRs with circular pools[9], the reactor container radial location was offset from the pool centre by about 0.7 m for SLOWPOKE reactors and by about 0.4 m for MNSRs to facilitate this removal (see Annex). The SLOWPOKE-1 prototype at Chalk River, SDR Whiteshell, PARR-2 and the University of Alberta SLOWPOKE-2 had pools that provided clearance distances larger than 0.7 m outside the reactor container.

The offset location of the container allows a spent active fuel cage to be loaded into a transport flask at the bottom of the pool from where the flask can be removed and placed inside an overpack container for shipment (see Annex). Neither SLOWPOKE reactors nor MNSRs were designed to be refuelled. However, the simple container design has nevertheless enabled the refuelling of eight reactors to date.

---

[9] PARR-2 and University of Alberta had rectangular pools (see Annex).

For SLOWPOKEs, both sections of the reactor container are fabricated from a 1 cm thick aluminium plate. The lower 82 cm long section has a 10 mm thick torispherical end. A ring on which the critical assembly sits is welded inside the lower torispherical end. The top plate on the upper container is 4 cm thick aluminium. For MNSRs, the lower section is 90 cm long with a thickness of 1 cm. The top plate on the upper container is 2 cm thick.

The lower and upper reactor container sections are held together by 16 aluminium tie rods of 1.9 cm diameter for SLOWPOKE reactors and 2.2 cm diameter for MNSRs. The rods are 4.6 m long and extend from the lower section flange to a square reactor container top support flange near the top of the upper section. The tie rods pass through slots in the lower section flange, with their heads located on the underside of the lower flange. The upper ends of the tie rods are threaded and are secured by nuts to the square flange. To separate the lower and upper sections, these nuts are loosened and the tie rods are swung out of the slots in the lower section flange, away from the reactor container. When released, the lower section sits 2.5 cm below the upper section and is supported by four arms, which extend radially outwards to hangers. The hangers are connected to the reactor support structure.

The reactor container is filled with water to a normal height above the upper surface of the core of 4.4 m for SLOWPOKE reactors and 4.7 m for MNSRs. The volume of water in the container up to this water level is $\approx$ 1300–1500 L, with allowance for the in-container component volumes. This leaves a gas filled headspace of $\approx$100 L volume within the reactor container; this is a height of about 39 cm inside the top of the container. Make-up container water requirements have historically been very low for both SLOWPOKE reactors and MNSRs, as there are negligible evaporative losses. Many facilities have reported no make-up requirements during an annual reporting period. One facility experienced a low container water level requiring small make-up additions of <5 L owing to leaks from the external purification system components when operating.

The reactor container has a support structure suspended from a steel angle frame resting on two steel I-beams placed across the pool (see Fig. 5). The reactor is suspended from the frame by four aluminium angle hangers, which are bolted to the square flange of the reactor container. The hangers are of such a length that the pool water and reactor water levels are kept at the same height above the upper surface of the core.

3.1.3.1. Reactor container gas purge system

Gaseous and volatile radionuclides released from the reactor water are retained in the enclosed reactor container gas space above the reactor container water surface and are vented to the reactor room ventilation system by a reactor container gas purge system. The system has an air supply, water trap and valving. The gas space is air purged to the reactor room ventilation system exhaust at the start of each working week. This limits the potential accumulation of hydrogen produced by the radiolytic decomposition of the reactor container water. There are no ignition sources in the gas space, but it is considered necessary to ensure that the hydrogen concentration does not increase to a level with potential for hydrogen and oxygen ignition. The weekly purge, typically scheduled for Mondays but always carried out following a minimum shutdown period of at least 48 h, reduces the concentration of radionuclides released from the headspace gas purge by allowing the decay of short lived isotopes. The purge system increases the reactor container gauge pressure by a small amount (<1 kPa). A simple pressure relief tube prevents the reactor container gauge pressure from exceeding 1.5 kPa if a blockage occurs in the purge system exhaust filter.

The active gases in the reactor headspace have relatively short half-lives. When the reactor is not operating, the headspace overall activity can therefore decay fairly quickly. Variations in the amount of activity released then depend primarily on two factors: the extent of reactor usage during the week and the elapsed time between the last reactor use and the time of the release. Decades of operational experience have determined that this periodic purging is acceptable provided that the container water conductivity specifications are maintained.

3.1.3.2. *Reactor container water purification system*

The reactor container water is purified in a closed loop system to maintain the conductivity and pH specifications. The system minimizes the corrosion of the inside of the reactor container and its internal components, removes soluble and insoluble radionuclides (activation and fission products) and minimizes the radiolytic decomposition of the reactor container water. The system is not operated continuously but rather is usually operated for just a few hours at the beginning of weekly reactor operation, when the reactor is shut down. This intermittent operating mode is intended to mitigate any potential leakage of active reactor container water from a purification system piping leak into the reactor room when the reactor is not attended. As with the pool water purification system, the reactor container water purification systems vary between facilities. A pumped flow rate of $\approx$ 10 L/min is typical when in service, through ion exchange resins and possibly filters returning water to the reactor container. The suction end of the system inlet line is positioned $\approx$ 15 cm below the reactor container water surface so that in the event of a leak the loss of reactor water would be limited. The discharge of the system return line is around the level of the reactor critical assembly.

The reactor water level is periodically maintained by adding some pool water drawn off from the pool purification system and added to the container via the purification system make-up waterline. Use of pool water for make-up then minimizes the load on the ion exchange bed. An outlet line is provided for removing water samples from the system.

3.1.4. Reactor critical assembly

The safety design philosophy (see Section 3.9) of a limited maximum excess reactivity requires that the components of the reactor critical assembly (i.e. fuel cage and fuel elements, beryllium reflectors and supporting components) cannot be rearranged into a more reactive configuration. These components are therefore assembled into a fixed geometry to form the critical assembly. The assembly, except for the top beryllium reflector shims, is installed in the reactor container during construction. Once installed, the assembly geometry remains unchanged, except for the top beryllium reflector shim configuration, which is rearranged and added to, compensating for the long term reactivity decrease of the core.

A vertical section through the critical assembly (see Fig. 4) illustrates how the various components are assembled. An aluminium tray for the upper beryllium reflector shims is attached to the critical assembly top plate by a bayonet fitting. This fitting prevents the tray from moving horizontally or vertically. A tube in the centre of the tray guides the central control rod (see Figs 4 and 5) into the reactor core fuel cage assembly (for either HEU or LEU). Sometimes referred to as a birdcage, the core fuel cage comprises lower and upper flanges and tie rods for supporting the fuel elements and rests on the lower beryllium reflector. An annular beryllium reflector surrounds the fuel cage, with its vertical and horizontal movement restricted. The critical assembly is set on an aluminium platform bolted to a ring in the reactor container lower section [19]. Natural circulation coolant water enters the core through an inlet orifice gap between the lower beryllium reflector and the bottom of the beryllium annulus and exits through an outlet orifice gap between the top of the beryllium annulus and the lower shim tray (see Annex).

The SLOWPOKE and MNSR fuel cages are, respectively, 22 cm and 23 cm diameter spool-shaped assemblies made from aluminium (for HEU) or Zircaloy-4 (for LEU) (Fig. 6). The MNSR HEU and LEU fuel cage geometry and construction is similar, but not identical, to the SLOWPOKE geometry and construction. The cages, with heights of 22.8 cm for SLOWPOKE reactors and up to 24.8 cm for MNSRs,[10] have a central tubular spindle to guide and allow movement of the central control rod inside the cage.

---

[10] The MNSR-IAE fuel element was 27 cm long.

The final fuel element cage loading for SLOWPOKE reactors is performed during commissioning activities, after which the loading arrangement is not changed. For MNSRs, the fuel cage is fully loaded prior to insertion in the reactor. Special tooling is used to implement manual fuel cage lifting and lowering.

3.1.4.1. *High enriched uranium fuel element cage loading*

The total number of HEU fuel positions available in the SLOWPOKE HEU fuel cage was 342 [19].[11] An additional 666 smaller holes in the flanges were provided to accommodate cooling water flow. The actual number of HEU fuel positions used varied between 295 and 325 for the various SLOWPOKE-1 and SLOWPOKE-2 reactors. Unused HEU fuel positions were left vacant.

The prototype MNSR-IAE fuel cage comprised 11 rings of fuel elements in a concentric circle arrangement. The total number of HEU fuel positions available in the MNSR-IAE HEU fuel cage was 411, of which 376 were occupied by HEU fuel elements and 35 by dummy aluminium elements to fill all the available positions [37]. The commercial MNSRs have a 10 ring fuel arrangement accommodating a maximum of 350 fuel elements in the cage. The actual number of HEU fuel elements used varied between 342 and 347 [27] in the commercial MNSRs.[12] Unused fuel positions, between 3 and 8, are filled with aluminium dummy elements.

3.1.4.2. *Low enriched uranium fuel element cage loading*

For the 2016 conversion of the MNSR-IAE, the 411 element prototype cage was used, of which 354 fuel positions were occupied by LEU fuel elements and 57 by dummy aluminium elements (see Annex). The remaining MNSRs used the 350 element cage design. The actual number of LEU fuel elements used for the 2017 LEU conversion of GHARR-1 was 335, with 15 dummy elements. The number of LEU fuel elements used for the 2018 LEU conversion of NIRR-1 was also 335 LEU with 15 dummy elements. The MNSR IHNI-1 used only 302 LEU elements for its first fuel charge [30].

The total number of LEU fuel positions available in the SLOWPOKE-2 LEU fuel cage is 342, the same as for the HEU cage. The LEU fuel and sheath diameters are 1.7% larger than those for the HEU, and the LEU fuel stack length

---

[11] The original Chalk River SLOWPOKE-1 prototype fuel cage consisted of an inner and an outer cage with a total of 325 fuel positions available and was fully loaded with 325 elements [16]. SLOWPOKE-1 HEU fuel also had slightly different dimensions from those of SLOWPOKE-2 HEU fuel (see Table 3 in Section 3.2).

[12] The MNSR-SD has 342 fuel pins (see Annex).

FIG. 6. *Top left: SLOWPOKE LEU empty fuel cage JM0001 conversion 2015. Top right: SLOWPOKE LEU fuel cage partially loaded with 128 elements; JM0001 conversion 2015 (courtesy of International Centre for Environmental and Nuclear Sciences, University of West Indies, Jamaica). Bottom left: MNSR-IEA LEU fuel cage with five tie rods, partially loaded with fuel. Bottom right: GHARR-1 LEU fuel cage with 350 fully loaded positions, 335 fuel elements and 15 dummy elements (courtesy of CIAE, China).*

is 3.2% (7 mm) longer than that for the HEU elements (see Table 3 in Section 3.2). These changes were made to compensate for the lower effectiveness of the LEU core reflectors due to the higher neutron absorption in the LEU core. The changes also maximized the fuel mass per unit volume by using some of the space above the upper flange of the fuel cage to enable the LEU elements to protrude out (see Fig. 6). Fewer LEU elements (196) were required for initial criticality compared with the 295 HEU elements required in the HEU core.

3.1.4.3. Instrument tubes

The critical assembly top plate provides access for two instrument guide tubes leading from the reactor container top plate to the top of the beryllium radial reflector. In SLOWPOKE reactors, a neutron SPFD is inserted into an instrument hole located just above the top of the radial reflector, through one of the instrument vertical guide tube holes at the midradial distance in the beryllium annulus. This midradial reflector location provides a high representative thermal flux. The SLOWPOKE-1 reactor was provided with two flux detectors — one for autocontrol and the other for backup monitoring — using a chart recorder [26]. In the MNSR, two miniature fission detectors are used in two similar locations in the radial beryllium reflector. MNSRs use four thermocouples inserted into four instrument guide tubes — two for inlet temperature monitoring and two for outlet temperature monitoring.

Dry, air filled irradiation tubes provide access to the neutron flux for sample irradiations (see Fig. 4). Both SLOWPOKE and MNSR types provide five small (inner) irradiation tubes, sealed against reactor container water ingress, each capable of accepting capsules 7 cm^3 in size. These tubes enter the reactor container through the top plate and terminate in the radial beryllium annulus. Vertical movement of these tubes and of the critical assembly is prevented by these bulkhead tube fittings. Provision is also made for five large outer irradiation tubes (for 27 cm^3 capsules) to be installed outside the periphery of the beryllium annulus in the reactor container water. All the outer tubes extend from sockets welded to the critical assembly platform to bulkhead fittings in the reactor container top plate. Not all reactors have installed all five large tubes in the available locations. The typical simplified horizontal midcore cross-sections of MNSR and SLOWPOKE fuel cores are shown in the on-line supplementary files (see Annex). Some SLOWPOKE and MNSR facilities had one or more small (inner) irradiation tubes installed in outer site locations.

3.1.5. Modifications to the original design

Minor historical modifications such as component replacement, upgrading of the reactor water or pool water purification, gas purge, process parameter monitoring and alarms and building service systems (process water, air supplies, electrical power, fire protection, HVAC and liquid drainage) are not discussed here. Historical modifications from the original concept are discussed in Section 2.2.5. Some modifications were not considered to be significant changes to the fundamental design and were not included in the list. These were the LEU conversions, the addition of a larger beryllium top shim in two SLOWPOKE reactors to extend fuel lifetime and the implementation of a digital control system.

No major generic modifications or changes to the main reactor structure inside the reactor container have been necessary for safety or production reliability. The Polytechnique Montréal SLOWPOKE reactor added five outer irradiation tubes during core conversion, but these tubes were anticipated and allowed for in the original design (see Annex). The only significant permanent modification to a SLOWPOKE reactor was the addition to the RMC reactor of a D_2O filled thermal column and a vertically oriented beam tube for radiography, which implied extracting a neutron beam above the pool water, an area without shielding (see Annex). The MNSR IHNI-1 was specially modified for BNCT medical treatments, as will be the proposed SUT MNSR (see Section 3.11).

3.1.6. Electrical power supplies

The off-site electrical power supply was traditionally considered adequate to safely supply all power requirements of SLOWPOKE and MNSR facilities. Loss of off-site power for an indeterminate time would not be expected to result in any significant safety concerns; after a number of days there would be a need to operate the pool and reactor container water purification systems and to purge hydrogen from the reactor container gas head space (see Section 3.8). Nevertheless, SLOWPOKE reactors were originally provided with limited uninterruptable battery backup power supplies for some instrumentation and control DC loads. Over time, various SLOWPOKE reactors and MNSRs have nevertheless upgraded their power supply requirements as research reactor international standards have evolved. Some facilities have installed diesel backup power supplies and have also upgraded battery power supplies to include an inverter power supply for uninterruptable AC loads for parameter monitoring and control.

3.2. FUEL DESIGNS

This section provides information on SLOWPOKE and MNSR fuel designs, including HEU and LEU fuels.

3.2.1. High enriched uranium

All SLOWPOKE HEU reactors, each originally fuelled with 93% ^{235}U (see Section 2.2.1.1), have now been decommissioned, other than the Polytechnique Montréal and Jamaica SLOWPOKE-2 reactors, which were converted to LEU in 1997 and 2015, respectively. The SLOWPOKE HEU fuel design is thus now obsolete (for details of the SLOWPOKE HEU design, see Table 3).

Four MNSRs are currently operating with their original 90.2% ^{235}U HEU fuel: SZ, PARR-2, ENTC and SRR-1. The MNSR HEU fuel had been successfully used in China's 125 MW High Flux Engineering Test Reactor and was chosen to be suitable for the MNSRs [15]. The MNSR-IAE HEU fuel element rods were about 24.8 cm long and 5.5 mm in diameter with aluminium alloy 303-1 cladding of 0.6 mm in thickness and aluminium plugs welded and sealed at both ends (see Fig. 7(a)). The lower tapered end plug of the element tightly self-locks with an interference fit into the lower fuel cage flange for stability. The upper end plug of the MNSR HEU fuel element is free to move through a positioning hole in the cage top flange. The MNSR HEU fuel had one year of thermal testing to simulate reactor conditions and physical stability. Vibration testing for magnitude 3 and 6 earthquakes demonstrated no adverse effects, as has subsequent reactor operation experience.

3.2.2. Low enriched uranium

LEU fuel was first provided in a SLOWPOKE reactor for the first fuel core of the RMC reactor in 1985 and subsequently for the Polytechnique Montréal SLOWPOKE reactor fuel conversion in 1997 and for the JM0001 fuel conversion in 2015. The fuel cage assemblies were similar to the HEU cores but were made of Zircaloy-4. The element locations in the LEU fuel cages are all essentially the same as in the HEU cores. The SLOWPOKE and MNSR LEU fuel element design is based on the reliable CANDU UO_2 fuel design but uses a smaller outside diameter.

For both the SLOWPOKE-2 LEU reactors, RMC and Polytechnique Montréal, a 198 LEU fuel element configuration was used. For the JM0001 reactor LEU conversion, 193 fuel elements were required [40]. LEU fuel requires about 40% more ^{235}U in total to provide the same reactivity as the HEU fuel because of the additional neutron absorption by ^{238}U. As UO_2 has a higher density than the HEU U–Al fuel, about 33% fewer LEU elements of the same dimensions are needed. There is also less neutron leakage in an LEU core than in an HEU core. As a result, the top beryllium reflector is worth less in the LEU core. The LEU top reflector overall worth was therefore increased to give the same value as the HEU core by increasing the fuel stack length by 7 mm compared with the HEU core [21], to 22.7 mm (see Table 3).

The SLOWPOKE LEU fuel elements (see Fig. 7(b)) are uranium oxide using 19.7 ± 0.2 wt % ^{235}U (see Table 3 for some of the main LEU fuel design data for comparison with HEU fuel design data). The pellets are solid right circular cylinders with a diameter of 4.2 mm, but without the dished ends and the graphite outer coating used for the much higher power rated CANDU fuel pellets. The pellet length-to-diameter ratio is nominally about 1.5, but this is chosen during manufacturing to optimize production capability and economy. The pellet-to-sheath diametrical clearance is chosen to facilitate fresh fuel pellet loading inside the sheaths during element fabrication, to accommodate any pellet expansion during burnup and to minimize any sheath strain. A minimum axial clearance space is also provided for fuel stack expansion.

The elements are sheathed in Zircaloy-4 tubes welded to Zircaloy-4 end caps. The lower end cap has a protruding smaller diameter split spigot end of 2.8 mm in the outer diameter. To load the fuel cage, each element is lowered into the cage through a hole in the top flange until the lower spigot engages a corresponding hole in the lower flange. The elements are then permanently secured in the cage's lower flange holes by peening over the element split spigot ends (see Fig. 7(b)). The element top end caps protrude slightly from the cage top flange and are free to thermally expand axially upwards through corresponding holes in the top flange. The LEU fuel cage has the same number of water cooling holes and fuel holes as the HEU cage. The CANDU fuel element design has the sheathing collapse onto the UO_2 pellets under the high pressure coolant conditions.

In SLOWPOKE reactors, the pellet–sheath contact will not be as firm as in a CANDU power reactor, as the reactor container cooling water is not pressurized. The pellet–sheath diametrical clearance is likely to decrease with burnup, as the pellets will expand slightly due to radiation induced swelling. Under SLOWPOKE conditions, the swelling will be very small — assuming natural fuel CANDU experience is also applicable to the low burnup SLOWPOKE LEU fuel. The void within the fuel sheath is filled with helium prior to end cap welding.

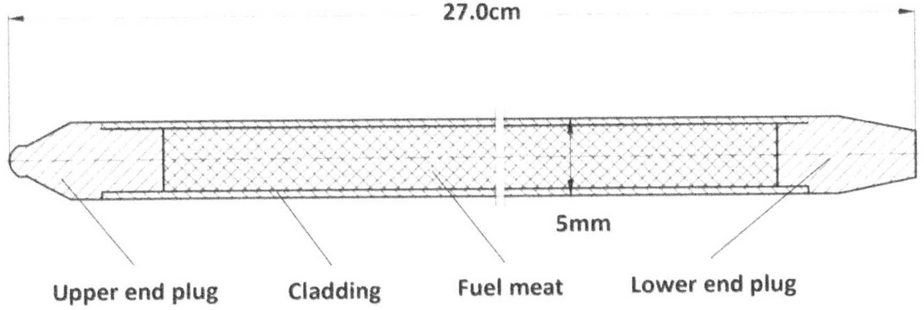

FIG. 7(a). MNSR-IAE HEU fuel element (courtesy of CIAE, China).

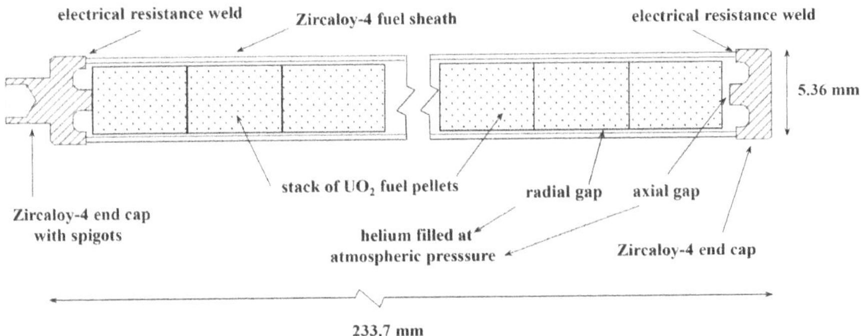

FIG. 7(b). SLOWPOKE LEU fuel element (courtesy of D.J. Winfield, Canada).

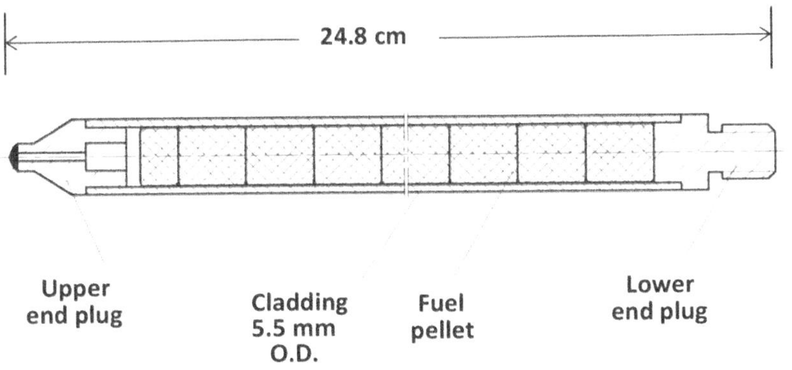

FIG. 7(c). MNSR LEU fuel element (courtesy of CIAE, China).

The presence of helium enables leak detection checking during fabrication and provides some improvement in pellet–sheath heat transfer and less potential corrosion compared with air. There is a negligible effect on core reactivity due to the helium filling from all the elements.

MNSR LEU fuel conversions were completed for MNSR-IAE in 2016, GHARR-1 in 2017 and NIRR-1 in 2018 (see Annex). The initial fuel charge for IHNI-1 was LEU. The LEU fuel cages and element locations remained essentially the same as in the commercial MNSR HEU fuel cages. The LEU fuel cage material is Zircaloy-4.

Tables 3 and 4 provide HEU and LEU fuel data for SLOWPOKE-2 reactors and MNSRs, respectively. The main difference between the SLOWPOKE and MNSR LEU fuel is the enrichment, which is around 19.75% in SLOWPOKE reactors compared with around 12.5% in MNSRs. The rationale for and consequences of using the lower enrichment in the MNSRs is discussed in Ref. [27]. Other differences are the dimensions of the LEU elements (see Figs 7(b) and 7(c) and Tables 3 and 4).

The pellet–sheath diametrical clearance is chosen to facilitate fresh fuel pellet loading inside the sheaths during element fabrication, to accommodate any pellet expansion during burnup[13] and to minimize any sheath strain. A small axial clearance space is also provided for fuel stack expansion. Figure 7(b) schematically shows the axial gap clearance at the top end cap of a SLOWPOKE LEU fuel element.

To date, the MNSR-IAE and Ghanaian and Nigerian MNSRs have been converted to LEU UO_2 pellet fuel (12.5% for the MNSR-IAE, 13.0 % for Ghana and Nigeria). Zircaloy-4 alloy is used for the cladding. The MNSR IHNI-1 used 12.4% LEU for its first fuel charge in 2009. The structural design of the MNSR LEU fuel element was changed from that used for the HEU fuel element. The LEU fuel element is placed into the grids of the fuel cage and the lower plug is fastened to the lower grid plate with a thread fitting. Compared with the conical lower plug fitting of the HEU elements, the LEU element thread fitting provides better structural stability. The cladding tube is girth welded to the upper and lower end plugs. The upper plug is welded closed after being filled with helium (see Fig. 7(c)).

---

[13] To date, no irradiated SLOWPOKE LEU or MNSR LEU fuel has been subject to a post-irradiation examination to confirm the sheath–pellet conditions after irradiation. Reference [38] provides results from a metallurgical examination of an unirradiated archived SLOWPOKE HEU fuel element as well as results from an in situ visual underwater television camera examination of the entire outer elements ring of fuel element and portions of some inner elements of the Polytechnique Montréal irradiated HEU fuel core.

TABLE 3. SLOWPOKE-2 HEU AND LEU FUEL DATA

Component	Parameter	LEU	HEU[a]
Fuel	Material	UO_2	U–Al alloy (28 wt% U)
	Enrichment (wt% of uranium)	19.75–19.89	93
	Element / pellet diameter (mm)	4.18	4.22
	Fuel stack length (cm)	22.7	22.0
	Density (g/cm^3)	10.6	3.45
	Equivalent boron concentration	<2	n.a.
Fuel sheath	Material	Zircaloy-4	Al
	Length (cm)	22.96	22.72
	Outside diameter (mm)	5.32	5.23
	Inside diameter (mm)	4.24	4.22
	Thickness (mm)	0.54	0.51
	Density (g/cm^3)	6.55	2.70
Core data	Volume of water in core, inside beryllium reflectors (L)	8.2	7.7
	Total mass of UO_2 (kg)	6.28	n.a.
	Total mass of U-235 (kg)	1.137	0.818
	Number of fuel elements	193–198 [b]	295–325 [c]

Note: Table 3 also includes data from Refs [22, 25, 38, 44, 45] and from the supplementary files (see Annex); n.a. – not applicable.

[a] SLOWPOKE-1 HEU fuel elements had slightly different dimensions [16].

[b] RMC SLOWPOKE required 198 LEU elements to achieve 3.46 mk excess reactivity (pool temperature of 33°C) with a 19 mm top beryllium reflector [41]. For 3.69 mk excess reactivity, 198 LEU elements were needed at Polytechnique Montréal [42] with the top reflector tray, no top reflector and a pool temperature of 33°C. JM0001 used 193 LEU elements for 3.8 mk excess reactivity, pool temperature of 33°C and one top reflector shim, 1.6 mm thick [43].

[c] See Refs [22, 38].

TABLE 4. COMMERCIAL MNSR HEU AND LEU FUEL DATA

Component	Parameter	LEU	HEU
Fuel	Material	UO_2	U–Al alloy (28 wt% U)
	Enrichment (wt % of uranium)	12.4–13.0	90.2-92
	Element/pellet diameter (mm)	4.0	4.0
	Fuel stack length (cm)	23.0	23.0
	Density (g/cm^3)	10.6	3.46
	Equivalent boron concentration	<1	n.a.
Fuel sheath	Material	Zircaloy-4	Al alloy 303-1
	Length (cm)	24.8	24.8
	Outside diameter (mm)	5.5	5.5
	Inside diameter (mm)	4.3	4.3
	Cladding thickness (mm)	0.6	0.6
	Density (g/cm^3)	6.55	3.46
Core data	Volume of water in core, inside beryllium reflectors (L)	8.2	7.7
	Total mass of UO_2 (kg)	33.3	
	Total mass of U-235 (kg)	1.298	0.914
	Number of fuel elements	302–335	342–347
	Number of dummy elements (aluminium)	48–15	8–3

Note: Table 4 includes data retrieved from Refs [25, 27, 37] and from the supplementary files (see Annex); n.a. — not applicable.

3.3. COOLANT AND MODERATOR

The critical assembly is immersed in the reactor container light water, which acts as a coolant and moderator and also as a radial, lower and upper reflector (see Fig. 4). The container coolant and moderator water extend vertically about 4.4 m above the core to provide adequate radiation shielding above the fuel.

3.4. BERYLLIUM REFLECTOR

SLOWPOKE reactors and MNSRs are light water moderated, low critical mass reactors characterized by their small physical size and high leakage of fast neutrons from the fuel core region. As a result, thermal neutron fluxes obtained inside a radial beryllium reflector can be higher than the lowest thermal flux in the fuel core (see Annex). A radial beryllium reflector is therefore used to accommodate sample irradiation tubes that are easily replaceable, outside the core in the high thermal flux beryllium reflector location. The bottom axial beryllium reflector also helps to reduce the fuel core size. The annular, bottom and top reflectors (see Fig. 4) are reactor grade beryllium metal.[14] The specific dimensions of these various reflectors vary between SLOWPOKE reactors and MNSRs (see Section 2.2.3). The beryllium top reflector is of variable thickness, as described in Section 2.2.3. Each semicircular plate has an outer diameter of 24.3 cm for both SLOWPOKE reactors and MNSRs. Different thicknesses — from 1.5 to 12 mm — are available to provide the different reactivity additions required as fuel burnup proceeds. The maximum top beryllium thickness that the shim tray will accommodate is 10.2 cm for SLOWPOKE reactors and 11 cm for MNSRs.

3.5. CONTROL AND SHUTDOWN SYSTEMS

This section provides information on SLOWPOKE and MNSR control and shutdown systems, including analogue control and monitoring systems, digital control systems, and control rod and drive assemblies.

---

[14] Reactor grate beryllium metal was chosen because accurate impurity specifications and manufactured concentrations for both SLOWPOKE and MNSR beryllium reflectors were needed to ensure minimum neutron absorption as well as for accurate Monte Carlo N-Particle (MCNP) reactivity calculations.

3.5.1. Analogue control

All SLOWPOKE reactors were originally provided with analogue control and monitoring systems. The SLOWPOKE-1 reactor that was transferred from CRL to the University of Toronto in 1971 had the original prototype analogue control and monitoring system. The SLOWPOKE-2 reactor installed at the University of Toronto in 1976 had a Mark 1 analogue control and monitoring system, as did the Dalhousie University, Polytechnique Montréal and University of Alberta reactors. The SLOWPOKE-2 reactors at the University of the West Indies and at the Saskatchewan Research Council (SRC) had a Mark 2 analogue system. The RMC SLOWPOKE had a Mark 3 analogue system.

The analogue Mark 3 SLOWPOKE reactor control console consisted of a reactor control panel, a radiation monitoring system and a panel monitoring auxiliary support system. The control system operates a single control rod and drive assembly responding to a cadmium SPFD linear power signal on automatic or manual control. Because of the reactor's self-limiting power excursion behaviour and its administratively controlled maximum excess reactivity of <4 mk,[15] the original safety and licensing process determined that it was not essential for the low level neutron flux to be measured and recorded during shutdown or startup (see Section 2.2.4). Low level log power neutronic measurements for control were therefore not implemented. This is unlike all other conventional research reactor control systems, in which reactor power is continually measured from very low neutron source power and is controlled and protected by a combination of control rod withdrawal speed and automatic neutronic shutdowns on low log power, high log rate power and/or high power level.

During startup, the SLOWPOKE analogue control algorithm fully withdraws the control rod at maximum speed, allowing the reactor power to increase exponentially up to the desired set point level. Consequently, for normal operation, the control system characteristics require that the chosen power level be achieved as soon as possible following the startup transient. The control rod position is displayed on a panel meter and monitored on an analogue chart recorder. The reactor core outlet temperature in the coolant upper outlet orifice is monitored continuously with a single thermocouple that is located in an instrument tube (see Fig. 4) and is supplied with auxiliary power during a loss of off-site electrical power supply.

---

[15] The original SLOWPOKE-2 excess reactivity allowed was 3.4 mk. This 3.4 mk was originally chosen to provide about a 50% margin from the maximum experimental excess reactivity of 6.48 mk inserted in the SLOWPOKE-1 transient tests [16]. In October 1995, the University of Toronto licence was changed to allow up to 4 mk excess. Other SLOWPOKE-2 reactors then subsequently also used 4 mk. MNSRs use 3.5–4.5 mk excess reactivity.

For reactor shutdown, the SLOWPOKE control system is designed to bring the reactor to a subcritical state. Rapid shutdown to a highly subcritical state is not required because of the reactor's limited maximum excess reactivity and limited likelihood of larger reactivity additions. Consequently, shutdown is achieved by driving in the control rod at the maximum insertion speed. In the event the control rod fails to shut down the reactor, the auxiliary shutdown system (ASDS) (see Section 3.5.4.1) is available to manually facilitate a shutdown if the reactor is in the attended operation mode.

Original MNSRs, except for IHNI-1, which has digital control, were all designed with a dual mode control system: manual control using an analogue control console and an independent autocontrol mode using a digital control system [11]. As noted by Zhou Yongmao in 1987 [11], no other research reactor type used this type of control, neither at that time nor subsequently. PARR-2 also operates in dual mode, but its autocontroller has been updated with an advanced graphical user interface to address malfunctioning of the original autocontroller (see Annex). The console control system is controlled by analogue signal inputs, controlling the measured neutron flux according to the required neutron flux set point by movement of the control rod. The computer control system converts the analogue inputs to digital signals and uses these digital signals to control the reactor. Currently, the PARR-2 console control system has not been digitized, but planning and preliminary work has been done.

A difference between the MNSR and SLOWPOKE control systems is that the MNSR control rod worth is somewhat larger than that of SLOWPOKE reactors. The MNSRs then have a larger shutdown margin[16] than do SLOWPOKE reactors (see Sections 3.5.4 and 3.6.1.8). Additionally, the SLOWPOKE control system does not have an automatic shutdown system. In contrast, the MNSR control system uses two automatic shutdown parameters — one for reactor overpower and the other for high core inlet–outlet temperature difference — to drop the control rod into the core by means of an electromagnetic clutch on the control rod drive cable (see Section 3.5.4). The MNSR flux measurement uses channel redundancy (1-out-of-2 trip logic), with two miniature fission chambers located in vertical holes in the annular reflector at the same radial distance as the inner irradiation sites (see Annex). The MNSR analogue control system also measures core temperature with two pairs of thermocouples to obtain the temperature difference between the inlet and outlet of the reactor core — one pair for the DCC system and the other pair for the manual analogue control system.

---

[16] Shutdown margin is defined in SSR-3, para. 6.151 [9]. To apply to SLOWPOKE reactors and MNSRs, the assumption is that the single control rod is fully inserted into the core.

3.5.2. Digital control

The first SLOWPOKE reactor that changed from analogue to digital control was RMC in 2001. Planning started as a graduate student thesis at RMC in 1999. The intent was to keep all the control and monitoring functions of the original Mark 3 analogue system and to improve on their functionality. Ontario Power Generation AECL software engineering standards were applied. The SLOWPOKE Integrated Reactor Control and Instrumentation System (SIRCIS) was then developed at RMC. The system underwent programme and integration testing using a hardware mock-up with in-house and third party validation and verification. SIRCIS then underwent performance testing and evaluations by licensed operators at RMC. With the commissioning plan approved by the CNSC, SIRCIS was installed in a two week period in June 2001. The main improvements were new visual displays of process parameters, particularly for control rod position. Improvements were also made to limit the maximum flux during control rod movement on startup. Improved performance and ease of operation of digital control was obtained with continuous indications, improved low flux power control and consistent excess reactivity measurements. A duplicate SIRCIS system is used as a simulator for testing spare parts and for future improvements. The JM0001 SLOWPOKE changed to digital control in 2023.

The MNSR computer autocontrol system functions are basically the same as those of the manual analogue control system. The main differences are that the display interface of all the important operating parameters has more features, is more detailed and can more easily display and archive trends. The signal outputs from the fission chambers are amplified and input to the computer data acquisition system card and, after a current–voltage conversion, are converted into a digital output. The control rod motor drive circuit uses the output from the data acquisition system card to operate the control rod. The reactor core inlet and outlet temperatures are monitored by thermocouples in the temperature measurement control circuit and converted to digital outputs. The computer then calculates the temperature difference between the inlet and outlet and compares the temperature difference with the temperature difference set point. When the temperature difference between inlet and outlet exceeds the set point, a shutdown signal is sent to the electromagnetic clutch on the control rod drive cable to disconnect power, shutting down the reactor as the rod drops into the core.

At the same time, the computer is also comparing the measured neutron flux density and the protection setting value of the neutron flux density. When the neutron flux density exceeds the set value, the computer will also send an overpower trip signal to the electromagnetic clutch of the actuator by the data acquisition system card to initiate a shutdown. Only one of the two automatic shutdown trigger sources is needed to initiate shutdown with 1-out-of-2 trip logic.

The electrical conductivity measurement circuit in the control box is mainly used to monitor the water quality of the pool water and reactor container water. The MNSR's computer control system has undergone three design upgrades to date. The current plan is to upgrade the control system of the NIRR-1 MNSR to digital.

3.5.3. Control rod and drive assembly

The SLOWPOKE central control rod is a cadmium tube — 0.8 mm thick and 25.4 cm long, with an outside diameter of 3 mm — enclosed within a central aluminium cylinder, with a 1 cm diameter, 7.5 cm long aluminium cylindrical spacer above and a 3.2 cm long aluminium cylindrical spacer below. The cadmium tube, enclosure cylinder and spacers are totally enclosed in a welded anodized aluminium tube with a 4.2 mm outer diameter and a 1.7 mm wall thickness. The complete rod, 41 cm total in length, can move inside the centre hole of the fuel cage with a 20 cm travel range, guided by the central tubular perforated guide tube (aluminium for HEU and Zircaloy-4 for LEU) (see Fig. 4) at the centre of the fuel cage. The lower end of the control rod is located just above the top of the bottom beryllium reflector. The reactivity worth of SLOWPOKE-2 HEU and LEU reactor control rods is ≈5.3 ± 0.3 mk [41, 42], with variations being attributable to fuel, fuel cage and irradiation tube differences. In its drive assembly, the SLOWPOKE control rod is suspended from a single turn drum by a stainless steel cable. A stainless steel weight located between the rod and the cable ensures that the cable is kept tight. The drive assembly is mounted on the reactor top plate and incorporates a bidirectional AC motor that drives the rod at a constant speed within the travel distance. A cam in the control drive assembly actuates a limit switch and relay at the upper end of the control rod travel, which turns off the motor power. A similar switch and relay operate at the lower end of the control rod travel.

In commercial MNSR units, the MNSR HEU central control rod absorber is a 26.6 cm long cadmium tube with a 3.9 mm outside diameter and a 1.9 mm inside diameter [27]. Located inside this cadmium tube is an aluminium rod with a 1.9 mm outside diameter. This absorber section is then enclosed within the lower end of a stainless steel tube with a 4.9 mm outside diameter and 0.5 mm in thickness, providing an overall control rod length of 45 cm. Aluminium spacer plugs, each 9 mm long, are inserted above and below the cadmium tube section. The control rod moves with a range of travel of 23–25 cm (covering all MNSRs) inside the central aluminium guide tube of the fuel cage. In the LEU cores, the central guide tube of the fuel cage is Zircaloy-4. The design reactivity worth of the MNSR control rod is ≈7 mk. Measured reactivity worths of MNSR HEU control rods are ≈6.4–8.0 mk. For MNSR LEU control rods, 5.7 mk is a typical value [27]. Unlike in SLOWPOKE reactors, the lower end of the control rod assembly penetrates a hole 2 cm in diameter in the bottom beryllium reflector and also has a guide tube above the fuel core.

The MNSR control rod is thus similar, but not identical, to the SLOWPOKE control rod. The main difference is that it also operates as an automatic shutdown system rod using an electromagnetic clutch, dropping the rod on loss of power when deactivated by either of two trip parameters. MNSRs also have two to four reactivity adjuster rods located in aluminium guide tubes just outside the core support structure and inside the reactor container (see Annex and Ref. [27]). The reactivity adjuster rods are used to adjust the initial core excess reactivity to the required level. These adjuster rods are not connected to the control rod drive system or the control system; they are fixed in place. For all MNSRs, except the IHNI-1, the adjuster rods are aluminium rods inside stainless steel clad tubing located within air filled (such as GHARR-1) or water filled (such as NIRR-1) aluminium guide tubes. Typical reactivity worths are $\approx$0.4 mk in the HEU and LEU cores [27]. The IHNI-1 reactivity adjusters use cadmium sleeves instead of stainless steel.

3.5.4. Reactor shutdown methods

As noted in Section 3.5.1, SLOWPOKE-2 reactors are not provided with an automatic shutdown system. Fully inserting the control rod in either manual or automatic control system operation is the normal method of reactor shutdown, with a shutdown margin of 1.0–2.0 mk. MNSRs have two automatic shutdown parameters, as noted in Section 3.5.1, although using the control system control rod to achieve automatic shutdown is not independent of the control system. The MNSR shutdown margin range is 2.1–2.6 mk.

Supplementary shutdown capability is also provided by two alternative methods of shutting down SLOWPOKE reactors and MNSRs. These supplementary shutdown methods are the ASDS and the remote shutdown system.

3.5.4.1. Auxiliary shutdown system

The purpose of the ASDS, used for both SLOWPOKE reactors and MNSRs, is to provide an alternative and independent manual method of reactor shutdown in the event of reactivity control events. The ASDS design basis requirement is to ensure that reactor shutdown can be achieved if any one of the following events occurs:

— The control rod remains in a fixed or stalled raised position and cannot be moved with the control system with the reactor at power.

— An unexpected power rise occurs, and the control system response is slow or ineffective.
— Any unexpected power rise occurs during beryllium shim addition operations or other core component maintenance operations.[17]
— For MNSRs, the automatic shutdown system fails to operate when required.

A fast, automatically initiated shutdown is not required. Failure of the control system, while operationally inconvenient, is not a safety hazard since the most serious consequence would be a safe self-limiting power excursion. It is nevertheless prudent to have a secondary means of achieving stable reactor shutdown in order to facilitate control system repair as soon as possible. The design basis requirement of the ASDS for all the events listed above is basically the same as achieving shutdown with the control system. ASDS operation is facilitated with a manually operated system for inserting cadmium-containing polyethylene capsules (ten capsules is typical) into, typically, all five inner irradiation sites (see Section 3.6.1.6).

Important safety features of the ASDS are that:

— The operation, maintenance and testing requirements of the ASDS are contained in the facility OLCs.
— The ASDS capsules have to be clearly identified and stored in an accessible location close to the capsule irradiation transfer system. Identifying the capsules minimizes the likelihood of the capsules being utilized for a routine NAA operation and possibly being misplaced. The system testing scheme is determined by facility-specific regulatory requirements.
— Periodic replacement of the ASDS polyethylene capsules is required to prevent embrittlement from radiation and ageing. Embrittlement could result in loss of capsule integrity and possible failure to complete insertion or possible entrapment of the absorber in the bottom of an irradiation tube. ASDS operability and, importantly, the reactivity worth of the cadmium capsules needs to be verified at least once per year. If the ASDS is inoperable or if any of the cadmium capsule vials are damaged, they need to be repaired before the reactor is operated.

Capsule transfer systems are normally operated using both off-site electrical power supply and compressed air, but in some facilities they can be operated

---

[17] Additionally, during the SLOWPOKE approach to first criticality procedures, the possibility may also arise of a power transient in the event of an element loading error. The SLOWPOKE approach to first critical philosophy itself is designed to provide protection against fuel cage criticality loading errors.

with compressed air alone. If both compressed air pressure and off-site electrical power supply are unavailable, the ASDS capsules can be inserted manually. This is achieved by pushing them into an opened capsule-loading station and then into each of the inner operational irradiation sites with a length of small diameter flexible polytetrafluoroethylene or polyethylene tubing.

3.5.4.2. Remote shutdown system

The remote shutdown system is also available in the event the reactor room, and hence the ASDS system, becomes inaccessible to the operators. The control system console and the control rod both still need to be operational for the remote shutdown system to operate. In practice, it is more likely that the remote shutdown system would be initiated (i) for some external or internal event in the reactor building where available time to access the control room is judged to be insufficient or (ii) when the control room is unattended while at power and is inaccessible for some reason. A manually operated reactor shutdown push button, located outside the control room, is usually provided. This push button then provides a remotely operated shutdown, which drives the control rod into the core via the control system console. This type of shutdown can be initiated by any reactor staff member or reactor user in the event of any type of emergency. Some reactors, such as PARR-2, may not have this remote shutdown feature installed.

3.5.4.3. Digital computer controlled shutdown

The SLOWPOKE and MNSR analogue control systems do not have an automatic shutdown system to drive the control rod in, in the event of a high flux condition or if the control rod remains in a fully up position. A shutdown can be achieved manually, overriding automatic control. This is done by use of a key switch that provides power to drive the control rod fully down in the event the control rod is movable. The RMC digital control is programmed to shut down the reactor using a control system rod insertion when either steady-state flux operation or the increasing neutron flux exceeds licensed limits (1.05×10^{12} n·cm^{-2}·s^{-1} and 1.40×10^{12} n·cm^{-2}·s^{-1}, respectively) (see Annex). The new JM0001 DCC system has a control algorithm that provides for a control system shutdown on high flux, high core outlet temperature, and high inlet and outlet core temperature difference. The MNSR digital control system does not provide a similar controlled shutdown algorithm, as the control rod magnetic clutch rod drop feature provides a rapid shutdown.

3.6. REACTOR PHYSICS AND THERMOHYDRAULICS

This section provides information on SLOWPOKE and MNSR physics, reactivity, neutronics and thermohydraulics, including safety limits.

3.6.1. Reactivity of system

In the following subsections, various contributors to core reactivity are discussed.

3.6.1.1. Temperature reactivity feedback

The dominant common reactivity related characteristic of LEU and HEU cores is under-moderation.[18] The most significant of the reactivity related mechanisms is then the moderator density temperature change. Reactivity feedback effects due to temperature changes are discussed for the LEU core for four main contributors: fuel, coolant water, beryllium reflector and reactor container downcomer water reflector; see (a) to (c) below. While the temperature reactivity effect of all these contributors can be independently calculated, the overall temperature change reactivity effect as a function of power is difficult to predict accurately. The calculation of the overall power coefficient[19] of the core is described in (d) below.

Historically, up to around 1985, the comparison of SLOWPOKE HEU and LEU predictions of the individual temperature coefficients and their overall variation with temperature used the Winfrith improved multigroup scheme code suite. The comparison with experimental isothermal temperature reactivity measurements was not very accurate [46]. Large uncertainties in these neutronics calculations were related to the characteristics of the small core geometry with high heterogeneity, the high neutron leakage and the temperature effect of the reactor container water around the beryllium reflector. Power coefficient predictions made in 1997 [47, 48] using the DONJON code [49] improved upon

---

[18] An under-moderated reactor has a negative moderator temperature reactivity coefficient. An over-moderated reactor has a positive moderator temperature reactivity coefficient. A power increase in an under-moderated core heats up the moderator, decreasing its density, with a net effect of decreasing neutron moderation causing a decrease in reactivity. A power increase in an over-moderated core also decreases moderator density, but the net effect is a decrease in absorption, causing an increase in reactivity.

[19] The change in reactivity from low power to the nominal full power condition is called the 'power coefficient', from power reactor terminology. This coefficient is the total reactivity change between these two conditions, rather than the reactivity change per °C used to describe the temperature coefficients.

the 1985 calculations. Subsequent 2017/2018 MCNP calculations to support SLOWPOKE [25] and MNSR [27] LEU conversion projects then provided the most updated, detailed and accurate calculations of temperature related reactivity effects for LEU and HEU reactors of both types. Only these most recent predictions are discussed below.

(a) Fuel temperature coefficient

MCNP5 calculations made in 2017 predicted a fuel temperature reactivity coefficient of −0.008 mk/°C [25] for the JM0001 SLOWPOKE LEU core, which is small and linear with power. The most recent 2018 MNSR MCNP fuel temperature coefficients are −0.009 mk/°C for LEU and +0.0004 mk/°C for HEU for average fuel temperatures between 127°C and 527°C [27].

For context, the predicted MNSR peak fuel temperatures during a rapid 4 mk reactivity excursion were 105°C for HEU and 148°C for LEU. The uncertainty in the calculated fuel temperature coefficient is smaller than the uncertainties in the calculated moderator and reflector temperature reactivity coefficients, because the fuel coefficient is not strongly influenced by neutron leakage. Direct and accurate measurement of the fuel temperature coefficient is not feasible practically, as fuel temperature measurements using instrumented fuel elements would be needed. Coupling effects from associated temperature changes of the adjacent coolant and reflector also limit calculational accuracy.

(b) Moderator and/or coolant temperature coefficient

As reactor power increases, fuel elements heat up and the local coolant water density in the core decreases. The reactivity of the coolant and/or moderator then decreases, mainly owing to the reduced moderation of the already under-moderated core. The overall effect is strongly negative and non-linear with coolant temperature and dominates in magnitude over other component temperature effects across the range of normal operating temperatures. The HEU coefficient is significantly larger than that for an LEU core.

For SLOWPOKE reactors, 2017 MCNP coolant temperature coefficient calculations for a core coolant ΔT of 22°C (about 20 kW power) gave −2.5 mk for the JM0001 HEU (93%) and −1.1 mk for the LEU (19.9%) [25]; MNSR coolant temperature coefficient calculations in 2018 for the HEU SRR-1 for a core coolant ΔT of 20°C gave −2.5 mk [27]. For MNSR LEU (13.0%) and the same core coolant ΔT of 20°C, the coolant temperature coefficient was −1.5 mk. Both of these values are very similar to those for SLOWPOKE reactors, despite the difference in LEU enrichments and LEU core arrangements. Direct comparison of the calculated moderator temperature coefficient through experiments is not feasible because of the coupling effects associated with temperature changes of the other components in the core.

(c)   Reflector temperature coefficient

The function of the neutron reflector for the core is performed not only by the beryllium reflector itself but also by the reactor container water outside the reflector. Both contributors affect the reflector reactivity temperature coefficient and are discussed below.

(i)   Beryllium

SLOWPOKE HEU temperature coefficient calculations up to 1985 did not consider any contribution from the beryllium reflectors [22], thus assuming a negligible effect. Later calculations made in 1997 with the DONJON code [47] confirmed this assumption and very small linear temperature coefficients for the combined bottom and annular beryllium reflectors (assuming no top beryllium shims) were calculated: +0.008 mk/°C for HEU (93%) and +0.002 mk/°C for LEU (19.8%). These values are mainly attributable to decreased absorption as the temperature increases. In later SLOWPOKE MCNP 2017 calculations, the HEU and LEU reflector temperature coefficients [25] were not recalculated owing to their small values and therefore low impact on reactivity safety. Subsequent MNSR MCNP 2018 calculations for the annular beryllium reflector only [27] predicted +0.002 mk/°C for HEU (90%) and +0.003 mk/°C for LEU (12.5%).

The reflector temperature coefficient uncertainty is expected to be larger than that of the other temperature coefficients because of uncertainties associated with the impurity composition and thermal expansion effects. The reflector coefficient is so small for both HEU and LEU that the uncertainty does not have a major effect.

(ii)   Reactor container water reflector

In addition to the beryllium reflectors (bottom, annular and top), the reactor container water just beyond the beryllium reflectors also acts as a reflector, mainly from the downcomer water inside the lower reactor container part of the cooling circuit outside the annular beryllium reflector. This coolant water container reactivity component, called for convenience the 'water reflector', was not addressed in the 1985 Winfrith improved multigroup scheme code suite calculations. This downcomer water reflector has a positive reactivity coefficient because it lies in an almost fully thermalized neutron spectrum region where further moderation is negligible and water absorption is reduced as temperature increases. The coolant water above the top of the core also acts as a reflector.

The 1997 DONJON SLOWPOKE calculations [47] gave an overall temperature coefficient for the combined downcomer water and top core water reflectors of +0.050 mk for HEU and +0.053 mk for LEU, up to 20 kW power. For the JM0001 LEU, the most current updated 2017 MCNP calculations [25] gave a much larger +1.5 mk for the water reflector reactivity worth for a core coolant ΔT of 20°C or 20 kW.

The 2018 MCNP MNSR calculations [27] then gave +0.009 mk for HEU and 0.01 mk for LEU water reflector reactivity worth for a core coolant ΔT of 20°C or 20 kW.

As with the moderator temperature coefficient, direct comparison of the water reflector temperature reactivity effects through experiments is not feasible because of the coupling effects associated with the temperature changes.

(d)   Overall temperature reactivity coefficient

The overall loss of core reactivity to full power is, for convenience, referred to as the power coefficient of reactivity. The measured LEU (19.75%) SLOWPOKE-2 power coefficient of reactivity (from low power 100 W to full power 20 kW) is about −1.2 mk. This is due to temperature induced reactivity changes from contributions (a), (b) and (c) above [25]. The close agreement of the experimental power reactivity measurements with calculation is then reasonable, considering the uncertainties of modelling the temperature reactivity behaviour of the various contributors. For MNSRs, section 2.10 of Ref. [27] calculates HEU and LEU (12.5%) power coefficient of reactivity values similar to those for the SLOWPOKE LEU core.

(e)   Isothermal temperature coefficient of reactivity measurements

Measurements of reactivity at low power were made in 1989 at both Polytechnique Montréal and RMC [50] to show the isothermal behaviour over an average core temperature range of 16–45°C. The core reactivity peaks at about 33°C (see Annex). The experimental data showed reasonable agreement with the calculations. It should be noted that the isothermal temperature measurements, while useful for analysis, do not reflect the time dependent reactivity change as power increases, as the various components all have different time constants with respect to power changes.

3.6.1.2. *Void reactivity effects*

Production of voids within the reactor core from boiling of the coolant and/or moderator displaces the moderator and reduces system reactivity. The École Polytechnique LEU void reactivity coefficient was experimentally simulated with a solid aluminium rod in the location of the central control rod [51]. A void reactivity of -1.43 mk was measured for the total volume of the central control rod position, 40 cm^3 or -0.036 mk/cm^3. This was equivalent to an overall core-averaged void coefficient of -2.8 mk/% core void, consistent with calculations of -2.6 mk/% core void, when corrected for the higher flux region used for the experiment.

For comparison with HEU, measurements made on the HEU SLOWPOKE-1 reactor core, using a 7 cm^3 void, gave a void reactivity of -3.3 mk/% core void at the central control rod location [19]. The HEU value was larger than the LEU calculated and measured core void coefficient owing to the larger coolant and/or fuel volume ratio of the SLOWPOKE LEU core, resulting in a less under-moderated core. The LEU void coefficient is still large, and void formation becomes important at high power for large reactivity insertions in providing negative reactivity feedback. For a given amount of thermal energy transferred from the fuel to the coolant, voiding produces a faster and much greater decrease in reactivity than moderator expansion does.

For MNSRs, the 2018 MCNP calculations for the void coefficient [27] gave -3.4 mk/% for HEU core void and the same value for LEU MNSRs (12.5%), which are values similar to those for SLOWPOKE reactors.

3.6.1.3. *Xenon reactivity effects*

During power operation, ^{135}Xe is produced in the fuel. The nominal full power equilibrium ^{135}Xe reactivity load of -2.6 mk would be reached after about 60 hours. About 9 hours after shutdown, the reactivity load reaches a peak value of -1.5 mk above this full power equilibrium value.

3.6.1.4. *Operational reactivity behaviour*

Assuming startup from a cold shutdown condition, with a pool ambient temperature of about 20°C, the above noted temperature effects collectively result in a typical overall core reactivity change $\Delta k_{core} \approx -0.24$ mk to reach a reactor power of 2 kW. To reach 20 kW in SLOWPOKE reactors and 30 kW for MNSRs, the Δk_{core} change is ≈ -1.2 mk. The average core ΔT is about 22°C at 20 kW full power. This is the same for the RMC LEU core as well as the

Polytechnique Montréal core for HEU and LEU [42]. MNSR HEU and LEU data give similar core ΔTs of 20°C and 26°C, respectively, for 30 kW full power [27].

On a longer timescale, ^{135}Xe poison reactivity loss becomes significant and the equilibrium ^{135}Xe load at full power adds about another −2.6 mk (see Section 3.6.1.3). The Polytechnique Montréal LEU core with 3.7 mk excess reactivity can then operate for about one day at full power (≈20 kW) with the reactor coolant water maintained at 33°C before this excess is lost owing to temperature and ^{135}Xe poisoning. This is an operational advantage for SLOWPOKE LEU fuel, as the JM0001 and Polytechnique Montréal HEU cores could operate for only 16 h at full power with 4 mk excess reactivity. At 3 mk excess reactivity, the maximum full power operational time for LEU is about 12 h. For HEU, this was only 6 h. As with SLOWPOKE reactors, MNSRs cannot be operated continuously at nominal full power because of ^{135}Xe buildup to equilibrium level. The full power operational duration of MNSRs is up to ≈4.5 h at 15 kW and up to ≈2.5 h at 30 kW [27]. The NIRR-1 LEU with 3.8 mk has operated for 5 h at full 30 kW power and for 8 h at 50% power. At 2.8 mk, these operating times dropped to 3.5 and 7 h at full power and 50% power, respectively [52]. A typical full operational schedule for MNSRs would be 2.5 h per day at full power, 4 days per week, 50 weeks per year.

As excess reactivity is reduced over time, the shorter term temperature and ^{135}Xe negative effects eventually balance the maximum positive excess reactivity of the system (i.e. the control rod becomes fully withdrawn) and the maximum power level can no longer be maintained. Additional reactivity can be obtained only by reducing the coolant temperature with a decrease in reactor power.

3.6.1.5. *Long term reactivity effects*

The main source of long term reactivity loss is from the stable isotope ^{149}Sm, which is mainly produced as the end product of the ^{149}Nd–^{149}Pm decay chain. The equilibrium level of ^{149}Sm is about −5.5 mk and is independent of flux and power level; ^{149}Sm is stable and decreases only by neutron capture. The rate of buildup to the ^{149}Sm equilibrium level is flux dependent and is very slow at the thermal neutron flux levels in SLOWPOKE reactors and MNSRs. The second major source of long term reactivity loss is consumption of ^{235}U by fission and radiative neutron capture. The third component of long term reactivity loss is the buildup of fission products with smaller cross-sections than ^{149}Sm. Core lifetime calculations for SLOWPOKE LEU (19.75%), MNSR HEU (90%) and MNSR LEU (12.5%) [27, 53] predict long term burnup reactivity depletions of −0.036, −0.049 and −0.039 mk/d respectively, assuming operation at nominal full power.

Operational experience from RMC, derived from RMC annual reports (2001–2005), using shim addition and accurate energy production data indicated a long term LEU burnup rate of −0.035 mk/d, assuming operation at nominal full power. This value is consistent with the calculated value.

3.6.1.6. *Auxiliary shutdown system reactivity characteristics*

In the event of a malfunction of the reactor control system, it is desirable to have an independent and alternative method of shutting down the reactor.

A manually operated cadmium capsule insertion system, the ASDS (see Section 3.5.4.1), is used for this purpose for both SLOWPOKE reactors and MNSRs. A minimum reactivity insertion of at least ≈ −4.8 mk is currently required for both SLOWPOKE reactors and MNSRs. The specific sequencing and loading or removal of the cadmium capsules, the reactivity worth of the ASDS capsules and the shutdown margin achieved can vary and depend on the existing capsule status of the irradiation sites at the time of usage. For SLOWPOKE reactors, experimental ASDS reactivity worth data using various numbers of cadmium capsules had not been well established before Polytechnique Montréal converted to LEU in 1997. Post-LEU commissioning tests at Polytechnique Montréal were performed in 1997 in order to establish an acceptable ASDS shutdown depth. These tests showed that an acceptable −5.1 mk was achievable for an LEU core. The reactivity characteristics of the ASDS can be predicted from calculations and easily confirmed with periodic testing during reactor operation. Reactivity measurements of varying numbers of cadmium capsules in various irradiation sites are typically found to be within 10% between different reactors, provided the ASDS cadmium capsule inserts have the same mass and shape. Overall reactivity measurements show that multicapsule reactivity worths are non-additive. The reactivity worth of a capsule will also change as the flux at the irradiation location varies with the number of top reflector shims and fuel burnup. Capsule reactivity worths are proportional to the square of the average thermal flux at their core location.

3.6.1.7. *Reactivity worth of top beryllium reflector plates*

An adjustment for the long term reactivity decrease is operationally desirable when excess reactivity has decreased by about ≈2 mk from the installed excess. Typical reactivity loss from the University of Toronto SLOWPOKE was −0.5 mk per year from an annual average energy usage of 15 MW·h, equivalent to operating at 10 kW average for 29 h per week. Typically, shim additions occur every 3–4 years in the LEU SLOWPOKE reactors. The JM0001 reactor can accommodate a maximum shim thickness of 10.2 cm. As core burnup proceeds and the beryllium top reflector thickness is correspondingly increased, the worth of the same beryllium thickness decreases. The top beryllium reflector incremental reactivity worth in the LEU core has been estimated using MCNP5 [25], showing that the shims (used in thicknesses from 1.6 mm to 10 mm) have reactivity worths ranging from 5.6 mk/cm to 0.29 mk/cm. The total worth of the 10.2 cm thick top beryllium reflector from the incremental additions predicted in Ref. [25] is about 20 mk, similar to that of the HEU core. The MNSR typical total worth of the top beryllium reflector is about 16 mk. Extended-life beryllium annulus additions (the big shim) for SLOWPOKEs are discussed in Section 2.2.3.

3.6.1.8. *Control rod reactivity characteristics*

The SLOWPOKE HEU JM0001 control rod had a reactivity worth of 5.4 mk [54, 55]. For comparison, the Polytechnique Montréal HEU rod worth was 5.18 mk [51]. The RMC LEU rod worth was 5.45 mk [41], the difference being attributable to the presence of the D_2O filled thermal column at RMC. As noted in Section 3.5.3, the MNSR control rod design worth is ≈ 7 mk. The worth of the control rod increases slightly and slowly over time because of changes in the axial flux shape due to the beryllium plate additions and burnup of ^{235}U relative to the rod cadmium. The effect is small compared with long term fuel reactivity decreases.

3.6.2. Thermohydraulic safety limits

CHF is the thermohydraulic safety limit that usually establishes the maximum allowed power level in commercial power reactors with forced circulation. However, in natural circulation SLOWPOKE reactors and MNSRs, the generation of steam in the core involves more complex flow phenomena than in forced circulation systems. It is then more difficult to estimate CHF accurately. The more conservative onset of nucleate boiling (ONB) power is therefore currently used as a more accurate and updated safety limit, even though heat transfer under natural circulation can improve significantly beyond ONB. Despite substantial research since the 1970s, particularly for natural circulation systems, large uncertainties still remain for predictions of severe overheating phenomena, as the heat transfer correlations are complex, involving many parameters. While heat transfer correlations have been developed for many research reactor fuel designs, none of these apply to the SLOWPOKE and MNSR HEU or LEU fuel cores.

The various thermohydraulic safety limits reached with increasing reactor power in a natural circulation system generally are:

— The ONB heat transfer stage, in which steam bubbles first form on the surface of the fuel cladding. Operating at ONB does not cause an operational problem in natural circulation reactors.
— The onset of significant void (OSV), in which steam bubbles break away from the fuel sheath surface when the void (steam) fraction is typically 5–10%. At OSV, some bubbles are released from the fuel element cladding, survive condensation and become entrained in the bulk fluid flow as the flow becomes two-phase.
— The onset of flow instability (OFI), a safety limit to be avoided, as sustained flow oscillations can cause forced mechanical vibration leading to physical damage of core components. Natural circulation systems, particularly in research reactors such as SLOWPOKEs and MNSRs, can become very unstable when significant void is present. Both OSV and OFI need to be avoided, as flow oscillations affect local heat transfer characteristics and can possibly induce burnout or hydraulically induced core damage.
— CHF, which occurs at a higher power and hence higher fuel sheath temperatures and heat fluxes than OSV. Full development of a steam film decreases the heat transfer coefficient through the steam void, raising the sheath temperature quickly, which will result in cladding failure due to burnout. CHF is sometimes referred to as the departure from nucleate boiling,[20] nucleate boiling being the regime in which an increasing steam (or void) fraction progressively improves cooling (or increases the heat flux between the fuel sheath and coolant). CHF is the local thermal power at which departure from nucleate boiling occurs.

Each of the above thermohydraulic safety limits are identified by the IAEA to be addressed for research reactor safety [56]. Sections 3.6.2.1–3.6.2.3 discuss how these safety limits have historically been addressed, as supporting calculations and experimental data have improved since the 1970s.

The original CRL prototype SLOWPOKE-1 HEU self-limiting power transient (1970) experiments [16] were performed up to a maximum of 6.48 mk, inserted in 0.8 s. The reactor attained a peak power of ≈186 kW and operated at >150 kW for ≈7 s. These transients were facilitated with the use of a curved absorber plate, worth about 4.5 mk, located close to the outer edge of the radial

---

[20] Various terminologies are used to denote the value of heat flux at the occurrence of CHF: dryout heat flux, burnout heat flux, maximum heat flux and departure from nucleate boiling heat flux.

beryllium reflector and operated manually using a motor-driven lead screw. Subsequent 1973 transient testing on the HEU SLOWPOKE-2 at Tunney's Pasture, Ottawa, up to a maximum of 6.05 mk, attained a peak power of 139 kW [16]. Two of the SLOWPOKE-1 and five of the SLOWPOKE-2 high power transients attained powers around 100 kW lasting for several minutes. The latter SLOWPOKE-2 reactor was also operated on normal automatic control at 80 kW for 49 min during experimental testing.

The prototype MNSR-IAE reactor maximum reactivity power transient from early testing was 3.6 mk, attaining a peak power of 76 kW. The MNSR IHNI-1 LEU core was also tested up to a 4.2 mk transient, attaining a peak power of 86 kW [30, 31]. Both of these MNSR transients were comparable with a Polytechnique Montréal LEU 4.3 mk commissioning transient of 77 kW [51] and also with a JM0001 LEU 4 mk transient of 63 kW [43].

During all of these original SLOWPOKE and MNSR high power transient tests, there was no indication of heat transfer instability or fuel damage. All transient tests proved to be demonstrably safe for the reactor and for personnel close to the reactor.

3.6.2.1. *1970s and 1980s safety limits*

Specific thermohydraulic safety limit acceptance criteria were not initially set for SLOWPOKEs and MNSRs. The strictly enforced excess reactivity limit and the rated maximum power level were the indirect criteria used to achieve an adequate safety margin with respect to thermohydraulic safety limits. The original 1969 CHF limit prediction used the Mirshak correlation, and for SLOWPOKE HEU the local CHF was predicted at ≈550 kW, a safety margin of a factor of ≈27 above the maximum power level [42].

The original 1985 thermohydraulics analysis for SLOWPOKE-2 LEU fuel, used for both RMC and Polytechnique Montréal, was based on codes and calculations described in Ref. [22].

3.6.2.2. *1990s safety limits*

The same 1969 CHF correlation was applied into the 1990s for LEU fuelled reactors at RMC and Polytechnique Montréal, the latter following its LEU conversion in 1995. Local LEU CHF was then predicted above a reactor power of ≈270 kW, a safety factor margin of ≈14 for a 20 kW maximum operating power. A conservative extrapolation of the peak power data for HEU and LEU transient peak power testing suggested that the LEU margin to CHF was thus still more than adequate for acceptable safety in an LEU core for power transients of 4 mk [22].

In preparation for the MNSR GHARR-1 and NIRR-1 LEU conversions in 2012–2018 (see Annex), the Argonne National Laboratory in the United States of America (USA) provided updated and supporting analysis estimates for various LEU and HEU thermohydraulic safety limits and safety margins in Refs [25, 57, 58]. ONB is thus now used as a safety limit for the SLOWPOKE reactors and MNSRs and is the most conservative thermohydraulic limit, setting the maximum allowable steady state power. ONB can be predicted more confidently than the other safety limits (e.g. CHF, OSV, OFI). In March 2016, the prototype MNSR-IAE core was converted to 12.5% LEU.

The power level at which ONB occurs was recalculated in 2018 for the HEU and LEU cores of the generic MNSRs using a one pin model in the PLTEMP/ANL V4.1 code [25] and in the RELAP5 V3.3 code [27] as 65–79 kW (HEU) and 68–81 kW (LEU). For comparison, the corresponding ONB prediction for the JM0001 HEU core using the PLTEMP/ANL code was 36 kW [25], with a resulting safety margin of about 1.8 above nominal full power. The LEU ONB power is lower than for the HEU, as the LEU core has a smaller fuel surface area and therefore a higher thermal heat flux, while coolant flow remains relatively unchanged. For HEU, ONB occurs at a power level of 90 kW and OSV occurs at a power level of ≈350 kW. At 300 kW and above, flow oscillations are predicted in RELAP5-3D. For LEU, ONB occurs at a power level of 62 kW and OSV occurs at a power level of 164 kW. Temperature and flow oscillations occur in RELAP5 V3.3 at power levels above OSV. At power levels above ONB, the reactor operates in the subcooled boiling regime until OSV occurs at a power level of approximately 145 kW. As power increased above the OSV limit, oscillations in coolant flow rates, coolant void fraction and pin temperatures were predicted. CHF would be reached at a power level above OSV. For MNSR NIRR-1 with nominal full power of 34 kW, ONB was calculated (in 2012) [58] to be 65 kW and 68 kW for HEU and LEU (348 elements of 12.5% ^{235}U) cores, respectively. At power levels above ONB, calculations also showed that the reactor would operate in the subcooled boiling regime, with OSV occurring at a power level of ≈145 kW. CHF is quoted to be at a power level far above OSV, but no value is provided because OSV and OFI are more limiting. For comparison, similar calculations for NIRR-1 were completed using the RELAP53D code [57] and flow oscillations were predicted for LEU at 300 kW with OSV at 350 kW and CHF at an unspecified higher level.

SLOWPOKE reactors at nominal full power operate well below the ONB condition. For the LEU core, ONB is predicted at 33.5 kW, indicating that the LEU reactor needs to have a safety margin of about 1.6 above nominal full power. Therefore, even though heat transfer may continue improving significantly beyond ONB under natural circulation conditions, experimental evidence shows that it becomes more difficult to make reasonably accurate predictions of the onset of the more significant safety limits of OFI and CHF as nucleate boiling increases, so the conservative ONB power is chosen as the effective safety limit.

Although LEU and HEU fuel have similar external dimensions, the average surface heat flux per element for full reactor power is higher for LEU fuel by a factor of about 1.55 owing to the smaller number of elements in the LEU core: 196 for LEU versus 296 for HEU (see Table 3) and $\approx$10% higher nominal power. The currently predicted LEU power form factor for JM0001 is 1.76, compared with the 1.37 predicted in the HEU core [27]. This results in a peak surface heat flux that is about 2.1 times higher in the LEU core than in the 296 element HEU core. The localized peak surface heat flux of the hottest element indicates a large margin from full power to CHF conditions. The Zircaloy-4 sheathing melting point is about 1760°C but oxidation can occur above 1200°C. The melting temperature of the HEU U–Al alloy is about 650°C and the melting temperature of the UO_2 LEU fuel is about 2840°C, the latter providing a much larger safety margin than the HEU fuel.

3.7. IRRADIATION FACILITIES

All SLOWPOKE reactors and MNSRs have five small inner irradiation sites in dry aluminium tubes (2.2 cm outside diameter). These tubes are uniformly located and vertically penetrate the beryllium annular reflector at a radial location close to the maximum thermal neutron flux in the annular reflector. Up to five large (outside diameter of 3.8 cm for SLOWPOKE reactors and 3.4 cm for MNSRs) or small outer sites can also be installed, as decided by the SLOWPOKE or MNSR facility. Outer sites are uniformly located in vertical positions, just external to the radial reflector in the moderator, to about the same insertion location height as the inner irradiation site tubes (see Fig. 4).

Two neutron irradiation tubes used for calibration of neutron detectors are located outside the MNSR container, in the reactor pool. The location of the irradiation sites in an MNSR is shown in the supplementary files (see Annex). The RMC SLOWPOKE has a full complement of five large outer irradiation sites and five small inner irradiation sites. It is also unique in having a radial heavy water thermal column and a graphite block illuminator, redirecting thermal neutrons upwards with a divergent neutron beam tube having a divergence of 8.5 degrees, used for neutron imaging (see Annex). The Polytechnique Montréal SLOWPOKE has three large outer sites and two small outer sites, installed during the 1997 conversion. The JM0001 SLOWPOKE has one large outer irradiation site installed. The MNSR IHNI-1 has two vertical neutron irradiation tubes and three horizontal neutron beams — a thermal neutron beam, an epithermal neutron beam for BNCT and a thermal neutron beam for PGA. The ENTC MNSR has two vertical neutron tubes for neutron radiography (see Section 2.2.5.9).

The irradiation systems comprise the irradiation tubes and associated pneumatic capsule transfer systems and are interlocked with the reactor control console so that capsules can be inserted and withdrawn only if the console control panel permits operation. As with the inner irradiation sites, the axial location for the larger capsules is also at the highest accessible axial thermal neutron flux location. The thermal neutron flux peak at the midcore location of inner irradiation sites is $\approx 10^{12}$ cm$^{-2}\cdot$s^{-1} at maximum rated power. At the outer positions, the full power flux peak is $\approx 0.5 \times 10^{11}$ cm$^{-2}\cdot$s^{-1} [54].

The capsule transfer systems are connected to the irradiation tubes with plastic tubing and use compressed air to propel capsules into and out of the irradiation sites. The irradiation and return air tube assemblies can be easily removed from the reactor for replacement, if required, in the event of a tube leak. The design also permits the facility to choose the number of irradiation tubes to be installed, without any significant effect on core components or reactivity.

Typical capsules used in the small inner irradiation tubes comprise a 7 cm^3 polyethylene cylinder with snap-top closures. Capsule sealing is achieved by either heat sealing or the addition of a second cap. Only polyethylene capsules are used. A maximum of three capsules is the limit for an irradiation site. Similar but larger 25–27 cm^3 polyethylene capsules are used in the large outer irradiation sites. In all facilities, it is also possible to insert the ASDS cadmium capsules manually from the capsule-loading stations. This procedure is used as a backup manual shutdown operational mode to the ASDS (see Section 3.5.4.1).

3.8. CHEMICAL CONTROL

Chemical control of the pool water and the reactor container water is necessary (see Sections 3.1.2.2 and 3.1.3.2) [9]. The pool water is purified with a pool water purification system to minimize corrosion on the outside of the reactor container. The pool water conductivity is typically maintained at < 10 µS/cm with a pH range between 5.5 and 7. Historically, there have been no reports of any corrosion issues with any of the SLOWPOKE-2 or MNSR aluminium reactor containers due to the pool water. The reactor container water is purified to minimize corrosion of the inside of the reactor container and of all other components inside the container. A dedicated closed loop purification system for the reactor container water is used. Traditionally, all SLOWPOKE-2 and MNSR facilities only operate their container water purification system a few hours per week, while the reactor is attended. This intermittent operating mode is intended to prevent any potential leakage of active reactor container water from a purification system piping leak into the reactor room when the reactor is not attended. The reactor container water conductivity needs to be maintained at <3 μS/cm. A pH range between 5.5 and 6.5 has been historically maintained for all SLOWPOKE-2 reactors and MNSRs.

Maintaining high purity for the reactor container water is particularly important to minimize the production of hydrogen and oxygen gases from the radiolytic decomposition of reactor container water and from oxidation of the aluminium container surfaces. The latter is of importance only during the first few months of reactor operation [19, 42]. The radiolytic gases migrate into the gas headspace above the container water. The reactor container gas purge system is used to ensure that potential explosive concentrations of hydrogen do not accumulate in the container gas headspace (see Section 3.1.3.1).

3.9. SAFETY DESIGN PHILOSOPHY

The dominant SLOWPOKE safety design philosophy has been to limit the magnitude and availability of excess reactivity by both design and administrative control. Combined with the inherent safety features of a large negative moderator reactivity temperature coefficient and an overall negative power coefficient, potential power transients are self-limiting to safe levels, without the provision of a rapid automatic shutdown system. As a consequence of the 4 mk excess reactivity limit, a rapid insertion of this amount is not hazardous. It allows for some margin of uncertainty, for example due to accidental water filling of irradiation tubes or inadvertent NAA sample insertions of fissile material. This safety philosophy then enables the use of just one single control rod and one SPFD. The traditional, more complicated multidecade redundant control system neutronics is not needed. The MNSR design uses the same inherent safety features as the SLOWPOKE design. However, MNSRs are provided with an automatic shutdown system using a single control rod, but with two trip parameters and multidecade redundant neutron flux measurements for the control system (see Section 3.11).

For both SLOWPOKE reactors and MNSRs, once the reactor container is fuelled, closed and sealed, and after initial commissioning, it is not essential for safety to have indication and control of neutron flux over the complete range from shutdown to full operating power (see Section 3.5.1). The sealed core design is also a key safety design principle. Only a licensed reactor engineer, under administrative control, may perform and supervise maintenance activities inside the reactor container.

As a consequence of the limit on excess reactivity, the shutdown margin in SLOWPOKE reactors and MNSRs is much smaller than in most other types of research reactor (see Section 3.5.4). Other research reactor types usually have a large available maximum excess reactivity, typically 20–40 mk or more.

3.10. LICENSING

The licence renewal period for Canadian SLOWPOKE reactors in the late 1990s was every 3 years. After 2000, this period was extended to 10 years, and since 2023 it has been 20 years for RMC and 10 years for Polytechnique Montréal. For JM0001, the licence period is 1 year. The first licences for MNSRs in China were all for 20 years and then subsequently for 5 years. For other MNSRs, the current licence periods vary: 1 year for SRR-1, 2 years for NIRR-1, 5 years for ENTC and GHARR-1, and 10 years for PARR-2.

The licenses for all HEU fuelled SLOWPOKE reactors were originally supported by a generic SAR produced by the vendor. A facility specific HEU SAR was produced for the University of Toronto SLOWPOKE in 1975 and a facility specific HEU SAR was produced for JM0001 in 1985. Currently, all three remaining LEU fuelled SLOWPOKEs are supported by facility specific SARs. All MNSR licences are supported by facility specific SARs.

The Jamaican JM0001 originally had no independent national nuclear regulator. All activities were self-regulated with approvals provided by the Director General of the International Centre for Environmental and Nuclear Sciences, the reactor owner. However, until the 2015 core conversion project began (see Annex), the same SLOWPOKE main licensing conditions imposed in Canada by the Canadian nuclear regulator had always been implemented (e.g. no operator access to open the reactor container, use of a vendor-qualified reactor engineer for control system modifications and periodic beryllium shim addition).[21] An internal facility safety committee provided oversight of the reactor safety. Historically, IAEA Integrated Safety Assessment of Research Reactors missions [59] had also been used periodically to provide independent safety assessments based on IAEA safety standards. In 2011, the Government of Jamaica, anticipating the 2015 core conversion, designated the Ministry of Industry, Investment and Commerce as the parent ministry to establish a radiation safety authority under the auspices of the Bureau of Standards Jamaica. A nuclear oversight committee was then established within the Hazardous Substances Regulatory Authority to act as the de facto nuclear regulator to review all

---

[21] The JM0001 reactor OLCs do not allow unattended operation with the reactor at power.

operational aspects of JM0001, ensuring that the reactor conversion would be conducted in accordance with IAEA guidelines.

After the first criticality in 1971, the University of Toronto SLOWPOKE-1 was claimed to be the first reactor in the world without a requirement for the presence of an operator for up to 4 hours during operation at power. The licence then specified the conditions under which remotely attended operation was allowed [60, 61]. Subsequently, this remotely attended period was increased and, since the 1980s, all Canadian SLOWPOKEs have been licensed to allow remotely attended operation with the reactor at power for up to 24 hours [62].

MNSR facility licensing in China is performed by the China Atomic Energy Authority regulatory agency, the NNSA, which oversees the development of nuclear energy in the China. The first MNSR (IAE) was also licensed to allow remotely attended operation at power [11]. In 2016, the CIAE submitted a safety analysis and environmental impact report to the NNSA. The NNSA then approved the prototype MNSR-IAE reactor operation licence for 5 years. In 2020, the Chinese regulator required all MNSRs in China to have a licensed operator in the facility control room when the reactors are operated at power. For licence renewal in 2021, the safety analysis, environmental impact and periodic safety inspection reports were again submitted to the NNSA to apply for continued operation for 5 years, which was approved.

All other MNSRs, other than the Thai MNSR, are owned by national nuclear organizations. The various regulatory authorities are the Nuclear Regulatory Authority of Ghana, the Iran Nuclear Regulatory Authority, the Nigerian Nuclear Regulatory Authority, the Pakistan Nuclear Regulatory Authority and the Atomic Energy Commission of Syria. IAEA Integrated Safety Assessment of Research Reactors missions were deployed to the ENTC MNSR (Islamic Republic of Iran) and GHARR-1 (Ghana) MNSRs.

3.10.1. Staff requirements and qualifications

Since the first licensing of the commercial Canadian SLOWPOKE reactors, licensed operational and maintenance staff have been required [7, 63]. Currently, Canadian SLOWPOKE reactors require a licensed operator, a licensed reactor engineer and a licensed technician. The latter two maintenance-type positions can be held by facility staff members or by a contractor employed by the vendor. Only the licensed reactor engineer is permitted to have access to and supervise maintenance activities inside the reactor container, with a work authorization from the reactor manager. The reactor engineer and the reactor technician, who are usually but not necessarily contract staff, are not mandated to be at the facility unless specific non-routine tasks are required.

The current SLOWPOKE facility licences do not explicitly mandate a minimum number of licensed operators. All three currently operational SLOWPOKE facilities have more than one licensed operator on staff: JM0001 with five, RMC with six and Polytechnique Montréal with two. The GHARR-2 and PARR-2 MNSRs require a minimum of two licensed operators on staff to operate their reactors; both facilities, though, have more (five and four, respectively) than the minimum required. For MNSRs in China, one licensed operator is the specified minimum. ENTC, SRR-1 and NIRR-1 do not have a specified minimum number of licensed personnel, but currently these facilities have 3, 4 and 15 licensed operators, respectively, on staff. Detailed qualification requirements for the licensed staff are specified in the various facility documents that support the licence (e.g. SAR, OLCs, training programme).

SLOWPOKE users authorized by the facility are listed in each annual report and can operate the NAA facilities under supervision of a licensed operator. Typically, Canadian facilities have appointed as many authorized users as is required for their experimental programmes. There may be a large number of authorized NAA users, with many typically from outside the facility. All MNSRs currently have somewhere between three and ten NAA users, some of whom are also qualified as licensed operators. The number of authorized users in all the reactors may vary substantially over the years.

For all SLOWPOKE and MNSR facilities, IAEA safety standards require that the relevant Member State regulatory authorities establish specific reference levels for personnel radiation exposure. For instance, historical dose experience in Canada did not mandate that a user or licensed operator be designated a nuclear energy worker (NEW)[22], the Canadian term for occupationally exposed workers in nuclear facilities or working with nuclear materials. Radiation doses have been kept well below the annual occupational limit for members of the public and, in most cases, below the sensitivity of the detection equipment used (see Section 6.2).

---

[22] A Canadian nuclear energy worker (NEW) is a person in an occupation who may receive a dose of radiation greater than the prescribed limit for the general public. As noted in Section 3.10.1, three SLOWPOKE facilities chose voluntarily to assign NEW status to some staff. JM0001 SLOWPOKE personnel, while not subject to a Canadian regulator, have always met international dose limits for the general public, with no historical instances of personal contamination recorded to date.

3.11. SLOWPOKE/MNSR DESIGN AND OPERATIONAL DIFFERENCES

There are a number of design differences between SLOWPOKE reactors and MNSRs and one main operational difference, listed below. The original nominal maximum power levels have not changed since the first reactors, other than an increase to 34 kW for MNSRs after LEU conversion. The currently planned 45 kW of the SUT MNSR is more significant. This feature of an unchanged order of magnitude maximum power level[23] contrasts with that of TRIGA reactors [1] as power levels increased significantly over the years, from 10 kW to 14 MW.

The main design differences between MNSRs and SLOWPOKE reactors are:

— MNSR LEU enrichment is 12.5–13% in the converted reactors and 19.75% for the planned SUT MNSR. This compares with 19.6–19.89% for SLOWPOKE LEU enrichments. The SLOWPOKE enrichment was chosen on the basis of the highest enrichment available under the Reduced Enrichment for Research and Test Reactors programme, to minimize the total number of LEU elements. This resulted in ≈198 elements in SLOWPOKE LEU reactors (see Table 3). As noted in Ref. [27], it was proposed to use the same number of LEU fuel elements as in the MNSR HEU cores, enabling a lower enrichment ≈13%. Using ≈19.75% enrichment was predicted to be too large to meet the maximum excess reactivity requirement of 4 mk for MNSRs operating at 34 kW, except for the 45 kW SUT MNSR.

— The thermal flux of 10^{12} cm$^{-2} \cdot$s^{-1} at the inner irradiation sites is the same for MNSR rated full power of 30 kW and 34 kW for HEU and LEU cores, respectively, but is achieved at 20 kW for SLOWPOKE LEU and HEU cores (nominal power values). There has been no detailed comparative study of the comparative flux and power ratios. This is not a straightforward task owing to the large number of potential parameters, with interdependence complicating such analysis. As part of the Argonne National Laboratory support for the fuel conversion of both reactor types [23, 25, 64], the most likely sources for the rated power discrepancies are following differences between SLOWPOKE reactors and MNSRs:

  • MNSRs use four reactor adjuster rods (see Section 3.5.3), resulting in more neutron absorption in the core periphery. This reduces the

---

23 Other than the first SLOWPOKE-1 at the University of Toronto (relocated from Chalk River in April 1971), which initially operated at 5 kW until being uprated to 20 kW in 1973.

neutron flux in the annular beryllium reflector.[24] SLOWPOKE reactors do not use any reactor adjuster rods.

- The MNSR lower beryllium reflector has a smaller thickness and outer diameter than that of the SLOWPOKE reactor and also has a 2 cm inner hole, which the SLOWPOKE beryllium reflector does not have.
- The MNSR has a larger number of fuel pins and uses some aluminium dummy elements in some of the empty lattice locations. Additionally, four or five fuel cage flange tie rods are used, compared with the three tie rods of the SLOWPOKE.

— Two miniature fission chambers are used in MNSRs for control, compared with one SPFD for SLOWPOKE reactors. The fission chambers provide some redundancy and have a larger detectable flux range response than the SPFDs, which cannot measure the flux level during initial stages of the startup.

— Automatic neutronic and temperature control rod drop trip parameters are used in MNSRs, versus no automatic shutdown system and no rod drop feature in the SLOWPOKEs.[25]

— Some MNSRs do not use a pool water cooling system, whereas all SLOWPOKEs have one installed (see Section 3.1.2.1).

— There are some basic differences in the type of control system, as summarized in Section 3.5.1.

The main operational difference between SLOWPOKE reactors and MNSRs is in the approach to criticality for fresh cores for either HEU or LEU and similarly for refuelling of LEU cores. The first approach to critical for both SLOWPOKE reactors and MNSRs differs from that of most other research reactors as a result of the small compact nature of the core components and also because the small compact fuel cage core was not designed to add or replace individual fuel elements. The main procedural steps for both SLOWPOKE reactors and MNSRs for removal of a spent HEU or LEU core and installation of a new LEU core are very briefly summarized below to illustrate differences in the procedures.

The SLOWPOKE first approach to critical technique utilizes the traditional method of inverse count rate subcritical multiplication. Importantly, it uses two temporary very low power continual wide-range neutron flux measurements as core reactivity is increased to ensure early detection of any unexpected increase

---

[24] The number of and detailed design of MNSR adjuster rods varies, with up to four being used for adjusting the excess reactivity to the allowable limit of 4 mk at initial fuelling. They are not part of the autocontrol system.

[25] IHNI-1has an auxiliary control rod and two adjusters [30].

in reactor power due to a potential core loading error. It is important to ensure that the procedural steps are performed in a specified sequence, to ensure the change in net core reactivity after each step is as predicted to maintain subcriticality of the core until the final net excess reactivity loading is achieved.

An approach to criticality, measuring the k_{eff} of the core during fuel loading, is made by plotting the inverse boron trifluoride proportional counter and/or ion chamber count rate against the number of fuel elements loaded in the fuel cage. About 100 fuel elements are initially manually loaded into the cage (see Annex). The fuel cage is then lowered manually into the core and the k_{eff} measured. The cage is then removed from the core. Fuel element additions continue, with typically ten more loading cycles, into and out of the core, until criticality and a maximum allowable core excess reactivity of 4 mk can be accurately predicted and achieved with a last fuel element loading. The maximum excess reactivity reference conditions for LEU are defined as fission product free, zero power, reactor container water temperature of about 33°C,[26] control rod fully withdrawn, all irradiation tubes empty and no top beryllium reflector shims.

With the reactor critical at a very low power, the beryllium upper reflector aluminium shim tray is installed and the small reactivity reduction measured. The control rod is then calibrated with reactor period measurements. The reactivity worth of the six ASDS cadmium shutdown capsules is then confirmed in various combinations in different irradiation sites, as well as with capsules filled with water to simulate moderator-flooded irradiation tubes. Power transients are then performed for reactivity excursions of 1, 2, 3 and 4 mk for comparison with the predicted behaviour of temperature reactivity feedback from other LEU cores as well as HEU cores. To achieve a 4 mk transient, the addition of a beryllium top shim plate may be necessary. Automatic startup using the control system is then performed with the SPFD. With SLOWPOKE reactors, the two low power commissioning neutron flux detectors are then removed, leaving the SPFD as the only remaining flux detector.

The MNSR initial core startup process is quite different from that of the SLOWPOKE reactor. The MNSR has the Zero Power Test Facility available at CIAE. This facility is used to calibrate the reactivity of fully loaded fresh HEU or LEU cores to confirm the reactivity design requirements, before fresh fuel elements are shipped to customers. The SLOWPOKE approach to critical fuel loading cycles is then not required. The MNSR cores calibrated in the Zero Power Test Facility are unloaded from the fuel cage before shipping to a facility.

---

[26] This reference temperature corresponds to the experimentally determined peak reactivity versus temperature curve. For both RMC and Polytechnique Montréal LEU fuel, this is about 33°C. For the HEU cores, the peak reactivity was about 15–19°C (see Ref. [25] and Annex).

For both MNSRs and SLOWPOKE reactors, LEU fuel elements are shipped to the facilities in batches for criticality safety, as was done with HEU elements in the past.

3.12. ADVANTAGES AND DISADVANTAGES

This section summarizes the various main advantages and disadvantages of some of the design and operational features of SLOWPOKEs and MNSRs.

3.12.1. Advantageous design features

The main design feature advantageous for safety is the limitation of the available excess reactivity. This feature provides safe self-limiting power excursion responses to large reactivity insertions from the negative reactivity feedback characteristic (see Section 2.2.4). Following fuel loading, the reactor container is bolted and sealed and, from this time on, the fuel loading of the fuel cage cannot be increased, decreased or rearranged. Adequate administrative controls to ensure reactivity safety still need to be implemented for irradiation of any fissile samples and also during beryllium shim additions. Accidental flooding of an irradiation tube with water can typically increase reactivity only up to $\approx$ +0.6 mk. This type of unanticipated reactivity addition is assessed to be within an acceptable safety margin. As a consequence of the limited excess reactivity, the control system design then requires only a single control rod (see Section 3.9), simplifying operation and minimizing costs.

The sealed reactor container design and separation of the reactor container water from the pool water results in the pool and container water providing sufficient radiation protection for close working access around the top of the reactor. Radiation doses to reactor staff and users were intended to be lower than those prescribed for members of the general public. Operational experience has demonstrated that this is achievable (see Section 6.2). Subgrade installation, conveniently used in most of the reactors, is an additional design advantage for radiation protection (see Section 3.1.2).

3.12.2. Advantageous operational features

Only one operator is needed to perform operations such as startup, shutdown and adjusting the power. It only takes about three minutes to reach full power from startup and the reactor can be operated at any time according to need. The ability to licence a reactor for remotely attended operation, with no licensed operator in attendance on-site, has been applied in Canada and in the MNSR-IAE prototype, and it may eventually be possible to apply this in other SLOWPOKE reactors and MNSRs. At the time licensing for remote attended operation was introduced, it was deemed to be a major breakthrough for cost and operational simplicity and remains unique to these reactors.

3.12.3. Disadvantages

The reactor power and neutron flux cannot be substantially increased without increasing the excess reactivity to compensate for ^{135}Xe buildup. Increased power availability would increase the efficiency of sample irradiation activity. The additional complexity of operations and higher radiation doses would then likely necessitate some staff being full time and, for SLOWPOKE reactors in Canada, then needing NEW status, increasing administrative costs.

Fuel and material testing programmes are not practical at the full power thermal neutron flux. Additionally, unlike larger power research reactors, SLOWPOKE reactors and MNSRs cannot operate continuously at nominal full power because of ^{135}Xe poison buildup.

There have been no fuel failures in SLOWPOKE reactors or MNSRs that have required a whole fuel cage assembly to be removed. A single significant fuel failure would be very expensive, requiring the provision of either a replacement core or possibly a refurbished core. Such a fuel replacement activity was not anticipated in the design and likely would not be practical, particularly if the facility did not have appropriate hot cell facilities.

Where sample NAA irradiation operations provide a substantial part of a facility's income, cheaper alternate chemical analytical techniques may over time displace some NAA activity. It is anticipated that there could always be a long term role for NAA to provide unique isotope specific NAA services, if the low operating cost of these reactors can be maintained as viable.

4. APPLICATIONS

Research reactors can be used for a broad range of applications. Reference [65] presents comprehensive information on applications of research reactors, including descriptions of the capabilities and fields of application of the techniques and the requirements for implementing each application at a research reactor. That information is not repeated here.

Some applications require a certain power or (thermal, epithermal or fast) flux level, which means that not all research reactors can be used for all applications. Table I–1 of Ref. [65] is a simplified research reactor capability matrix listing applications accessible for each order of magnitude of research reactor power. Accordingly, research reactors with powers of ≈100 kW (closest to SLOWPOKE reactors and MNSRs) have full capability in NAA, education and training, and BNCT, and some capability (e.g. for demonstration or research and development purposes) in PGA, radioisotope production, geochronology, gamma irradiation and testing of instruments. A few SLOWPOKE reactors and MNSRs have been modified to incorporate neutron beam tubes, needed for BNCT, neutron radiography and PGA.

Table 5 shows the number of SLOWPOKE reactors and MNSRs involved in each utilization area, according to the RRDB, complemented by information provided in the facility reports in the supplementary files (see Annex). Information from decommissioned reactors was used where available. For clarity, NAA, PGA and delayed neutron counting (DNC), which are all included in activation analysis in the RRDB, are listed separately in the table. Two categories additional to those considered in the RRDB were introduced in Table 5: 'Reactor physics and analysis' and 'Other irradiations'. The former includes determination of reactor parameters, modelling and engineering studies supporting reactor operation, and planned changes including conversion to LEU. The latter includes new reactor concepts and irradiation applications not well covered by the other categories, such as radiation processing of materials.

The most common applications of SLOWPOKE reactors and MNSRs are NAA and education and training, with 10 or 11 operational and 6 or 7 decommissioned reactors being involved in these. These applications are followed by radioisotope production, with a total of nine reactors involved, of which five are currently operational. Other applications are implemented in smaller numbers of reactors. Geochronology planning is reported by one reactor. The RRDB notes for the MNSR IHNI-1, which was designed and used primarily for BNCT, that PGA and neutron radiography are also under development.

4.1. STRATEGIC PLANNING

Research reactors need to have effective and achievable strategic plans to support their long term sustainable utilization. Reference [66] provides information on how to develop a strategic plan for both existing and planned research reactors. One key aspect is to identify existing and potential stakeholders, and to assess and prioritize their needs in order to develop the facility's experimental capabilities to meet the users' needs. Table 5 shows that SLOWPOKE reactors and MNSRs have taken widely different paths regarding the mix of applications implemented, besides NAA and education and training.

TABLE 5. UTILIZATION OF SLOWPOKE REACTORS AND MNSRs

Application	Total	Under construction	Operational	Decommissioned
Education and training	18	1	11	6
NAA	18	1	10	7
PGA	3	1	2	—
DNC	4[a]	—	1	3
Radioisotope production	9	—	5	4
Geochronology	1[b]	—	1[b]	—
Neutron imaging	4	1	3	—
BNCT	2	1	1	—
Testing of instrumentation and control	3	—	2	1
Nuclear data measurements	1	—	1	—
Reactor physics and analysis	6	—	6	—
Other irradiations	4	—	3	1
Heating	1	—	—	1

Note: BNCT — boron neutron capture therapy; DNC — delayed neutron counting; NAA — neutron activation analysis; PGA — prompt gamma ray analysis.

[a] Besides the RMC SLOWPOKE, the Tunney's Pasture, SRC and University of Alberta SLOWPOKE reactors had DNC systems until decommissioning.

[b] The RRDB indicates that the JM0001 and NG0001 reactors have activity in geochronology. However, for the JM0001 reactor, this is in fact geochemistry and geological mapping with NAA, and this was planned for the NG0001 reactor but not yet implemented.

Some reactors, such as the Halifax and Saskatchewan SLOWPOKE reactors, were dedicated solely to those two applications, according to the RRDB.

Reactors such as the SLOWPOKE-1 and SLOWPOKE-2 at the University of Toronto (see Annex), the SLOWPOKE-2s at Tunney's Pasture, the Kanata Isotope Production Facility in Ottawa, the MNSR-SD and the MNSR-SH (see Annex) were shut down and decommissioned when their original purposes were fulfilled. At the time that this earliest group of reactors began operation, strategic planning for future decades of reactor operation was not a major consideration, and IAEA guidance on the subject did not yet exist. Other reactors, such as the Polytechnique Montréal SLOWPOKE-2 (see Annex), adapted their strategy over the decades according to the evolving needs of stakeholders, starting with those of upper management and of Canada's federal government. This led to a reorientation of funding towards self-sustainability. Searching for and finding new users, both in NAA and in new areas of application such as radiotracers, made the reactor's strategic plan more adaptable to change.

In general, long term sustainable utilization is currently a key consideration for both SLOWPOKE reactors and MNSRs, for which developing and maintaining an updated strategic plan is essential. For these reactors, typically built to develop capacity in nuclear science and technology, sometimes as the first nuclear installation in the country, all the strategic planning considerations in section 3.10 of Ref. [66] would still be applicable. Of these, stakeholder and user relations, marketing and financial management are identified in Ref. [66] as three dominant strategic considerations to justify continued operation or to build a new research reactor facility. IAEA Integrated Research Reactor Utilization Review missions [67] can be effectively used to assist operating organizations in their decision making process and, eventually, in the development of a strategy towards sustainable operation.[27]

4.2. EDUCATION AND TRAINING

Education and training is the most common application found at research reactors worldwide, with about two thirds of operational reactors engaged in some form of education and training. While education and training are separate disciplines, they use the same or very similar methods, instruments and experimental equipment. Education is directed at students in an academic context, often leading to BSc, MSc or PhD degrees, while training is directed at preparing nuclear professionals for specific activities.

Most of the SLOWPOKE and MNSR facilities, many of them located at universities or educational institutions, have been involved in providing some form of nuclear education and training. These reactors provide a convenient facility for education and training in a whole range of nuclear topics, such as radiation and reactor physics, thermohydraulics, nuclear materials, radioisotope production and radioanalytical chemistry, as well as instrumentation and control. They also offer educational opportunities through involvement in the various nuclear applications (see Sections 4.3–4.10). The simplicity of the small scale of operations provides advantages for hands-on activities that would be much more difficult to achieve using larger research reactors.

In 1975 and 1977, AECL Commercial Products Division, Ottawa, provided a nuclear engineering training package and NAA, DNC, uranium geochemical analysis, gamma irradiation and nuclear material absorption analysis manuals for

---

[27] See: https://www.iaea.org/services/review-missions/irrur

use in SLOWPOKE reactors. This had a significant historical impact on their use and acceptance in the university context [68].

Several SLOWPOKE reactors and MNSRs also provide tours for the public, which plays a positive role in increasing the public's knowledge about and acceptance of nuclear science and technology. For example, the MNSR-SZ receives more than 100 visitors yearly from a variety of backgrounds. The University of Alberta SLOWPOKE Facility regularly had more than 60 high school physics students from different schools tour the facility each year.

4.3. NEUTRON ACTIVATION ANALYSIS

SLOWPOKE reactors and MNSRs have been, and continue to be, extensively used for NAA. Of the reactors for which information on the utilization programme is available, only the SDR reactor and the IHNI-1 MNSR were not involved with any NAA activity.

A major advantage of NAA is the simultaneous non-destructive determination of a number of elements in almost any type of sample matrix including large bulk samples, without the need for special sample preparation [69]. Sensitivities are often at the mg/kg level (parts per million) level and can reach µg/kg (parts per billion) in favourable cases. NAA is amenable to automation of sample changing, irradiation and counting [70], allowing for the measurement of very large numbers of samples (see Annex).

The high sensitivity available has led to the widespread use of NAA in environmental, health and cultural heritage studies. One specific application of NAA is geochemistry, where geological formations can be studied. Ghana, Jamaica and Nigeria all developed national programmes for geochemical analysis. The JM0001 reactor in Jamaica has conducted extensive national and international collaborative research on this topic. Nationally, this research was used to create a Geochemical Atlas of Jamaica (see Annex). This capability was later used in forensics analysis [71, 72]. Similarly, NIRR-1 has also used NAA for geochemistry research in Nigeria (see Annex) and has also implemented k_0-NAA calibration [73]. The MNSR-SD reactor was also used in geochemical mapping, including a geological survey of the lower reaches of the Yellow River, trace elements in carbonate rocks, sediment samples in the Shandong Province, deep penetrating gold deposits, and rare earth elements in geological samples in Shandong (see Annex).

SLOWPOKE reactors and MNSRs are ideally suited for the provision of a commercial NAA service because they do not require long maintenance shutdown periods, making it possible to offer a reliable long term service. Uranium geochemical analysis and assay was a significant SLOWPOKE commercial activity in the 1970s [74]. The Montréal SLOWPOKE has an extensive NAA commercial service programme that allows for financial self-sustainability of the reactor operation, with major users such as manufacturers of plastic, fuel cell power products, and lumber and wallboard products, and an oil refinery, in many cases for quality assurance purposes.

As an example of the NAA sample irradiation productivity that can be achieved using short irradiation, decay and counting times [75], the University of Toronto SLOWPOKE has averaged 17 165 documented sample irradiations per year in its 22 year lifetime (see Annex). The University of Toronto SLOWPOKE reported a maximum number of irradiations in one day of 314 samples on 7 October 1996 [76].

The RMC SLOWPOKE and the ENTC MNSR report the use of low level counting of environmental samples using the gamma spectroscopy equipment available in the reactor facility, as did the University of Alberta SLOWPOKE until it was decommissioned.

4.4. PROMPT GAMMA RAY ANALYSIS

PGA is a specialized application of NAA, capable of determining elements not easily accessible with other techniques, including boron and hydrogen, and high cross-section elements such as boron, cadmium, gadolinium and samarium. It requires the availability of a neutron beam and is generally found at research reactors with higher power levels. PGA is therefore not commonly implemented at SLOWPOKE reactors and MNSRs. The ENTC MNSR is the only reactor in which it has been implemented, in a vertical beam tube. A second PGA facility is under development at the MNSR IHNI-1. The Thailand SUT MNSR also plans to implement PGA.

4.5. DELAYED NEUTRON COUNTING

DNC is similar to PGA, a specialized application of NAA. It is well suited for the determination of the uranium content in samples, which usually come from the mining industry. A DNC system has been used at the RMC SLOWPOKE for academic purposes (see Annex). The Tunney's Pasture, SRC and University of Alberta SLOWPOKE reactors had DNC systems. SRC used their system to analyse several tens of thousands of samples for the uranium exploration, mining and milling industries. The University of Alberta DNC system was largely used for geochemical and environmental purposes.

4.6. RADIOISOTOPE PRODUCTION

Radioisotope production is the third most common application of SLOWPOKE reactors and MNSRs, beginning originally in 1984 with the dedicated Kanata Isotope Production Facility SLOWPOKE in Kanata, Ottawa. Given the restrictions in flux and operational time at full power, production of radioisotopes is limited to those with a half-life from minutes to a few days. Some radioisotopes with a half-life up to a few months can be produced, albeit with low activity. The radioisotopes produced are used mainly in isotope labelling and tracer technology for research, industrial and medical applications.

At the University of Toronto SLOWPOKE-1 and SLOWPOKE-2 reactors, radioisotopes such as ^{18}F, ^{24}Na, ^{35}S, ^{43}K, ^{64}Cu (both carrier-free from Zn and from activation of natural copper), ^{115m}Cd, ^{128}I, ^{141}Ce and ^{185}W were produced (see Annex). At the RMC SLOWPOKE, ^{24}Na, ^{37}Ar and ^{140}La are produced (see Annex). A ^{46}Sc irradiator assembly for testing the performance of electronic components was designed at the Polytechnique Montréal SLOWPOKE using four ^{46}Sc neutron activated vials optimized to obtain quasi-homogeneous activity along the axis of the source (see Annex). Other SLOWPOKE reactors and MNSRs also continue to produce these and other radioisotopes. Over its operating history, the University of Alberta SLOWPOKE-2 reactor produced the following radionuclides: ^{18}F, ^{24}Na, ^{32}P, ^{38}Cl, ^{41}Ar, ^{42}K, ^{56}Mn, ^{59}Fe, ^{60}Co, ^{64}Cu, ^{82}Br, ^{140}La, ^{141}Ce, ^{147}Nd, ^{153}Sm, ^{153}Gd, ^{182}Ta, ^{187}W and ^{198}Au for research and/or industrial purposes.

4.7. NEUTRON RADIOGRAPHY

Neutron imaging can be performed on low power research reactors. Full 3-D tomography has been demonstrated at nearly zero power [77]; however, a neutron beam is required, which was not included in the original SLOWPOKE and MNSR designs. Nevertheless, 2-D neutron radiography has been implemented and widely used by making a significant design change on the RMC SLOWPOKE (see Section 3.1.5 and Annex) and on the ENTC MNSR (see Annex, Ref. [78] and Sections 2.2.5.7 and 2.2.5.9). In both cases, vertical beams were installed and used for neutron radiography experiments.

Neutron radiography is also under development at the IHNI-1 MNSR and planned for SUT MNSR (see Annex). In both cases, the reactor design has been significantly modified from previous MNSRs, with the introduction of horizontal beam channels dedicated for neutron beam applications.

4.8. BORON NEUTRON CAPTURE THERAPY

Research reactors have been used for BNCT for many years [79]. In recent years, several research reactor based BNCT facilities have shut down, and only a few reactors worldwide still engage in patient treatment, with some other facilities being used for research and development in BNCT. SLOWPOKE reactors and MNSRs are unique in that the reactor design can be modified to provide a dedicated BNCT facility that can be located in a medical facility. This takes advantage of the small size, low cost and simple reactor operation, which larger research reactors cannot provide. The reactor type is particularly advantageous, in terms of its low staffing and operational costs, as patient demand is not easily predictable.

The MNSR IHNI-1 reactor was built as a BNCT facility, with three horizontal neutron beams — two for epithermal neutrons and one for thermal neutrons — dedicated to BNCT. The reactor reached criticality in 2009 and full power in 2010. This was followed by clinical trials, with the first patients being enrolled in 2014 [80].

The SUT MNSR, under construction, has a design that evolved from the IHNI-1. In addition to the horizontal neutron beams to be dedicated to neutron imaging and PGA, it has one vertical neutron beam directed downwards to the BNCT room, which is located below the reactor core. It is expected to be ready for commissioning in 2026.

4.9. HEAT PRODUCTION

Initially, SLOWPOKE reactors were designed and intended for neutron production, but not for power or heat production. As noted in Section 3.1.2.1, the heat production rate in the pool water is very low, as expected from the low maximum reactor power rating and the separation of pool water from the reactor container water. There are therefore no useful practical applications of heat production from the pool or container water.

During the late 1980s, a project was developed to upgrade and dedicate the design for thermal low temperature power up to 2 MW(th) in a modified type of SLOWPOKE reactor (SDR), as described in Annex. Commissioning tests demonstrated the production of low temperature heat at temperatures below 100°C, from a small pool-type reactor design. However, the SDR reactor operation was short lived and other planned practical applications of air-conditioning and small scale electrical production were not pursued. A further design for a 10 MW(th) SLOWPOKE (SES-10) was also pursued, again with the prime focus on low temperature heat production, but this programme was terminated before the design was completed (see Annex).

4.10. NUCLEAR DATA

The low neutron fluxes available at the SLOWPOKE reactors and MNSRs are in general not conducive to extensive programmes for determination of nuclear data. The Nigerian NIRR-1 has been used to determine neutron spectrum averaged cross-sections for some light and medium mass nuclei, including the (n,p) reaction on ^{27}Al, ^{28}Si, ^{29}Si, ^{46}Ti, ^{47}Ti, ^{56}Fe and ^{58}Ni, and also for the (n,α) reaction on ^{30}Si [81].

5. HEU TO LEU CONVERSION AND LEU TO LEU REFUELLING

This section summarizes the history and activities required for HEU to LEU conversion and LEU to LEU refuelling for SLOWPOKE reactors and MNSRs.

5.1. HISTORY AND PURPOSE OF HEU TO LEU CONVERSION

To reduce the possibility of nuclear proliferation, J. Hilborn proposed LEU conversion to AECL for SLOWPOKE as early as 1977 [8]. In line with this proposal, in 1985 the RMC used LEU for its first fuel loading,) and in 1977 the first SLOWPOKE HEU to LEU conversion took place at Polytechnique Montréal in 1997. By 1997, the Polytechnique Montréal HEU core was approaching the point at which the addition of shims to the top beryllium reflector would no longer compensate for the loss of reactivity due to burnup after 31 years of operation. The next HEU to LEU conversion, for both non-proliferation and burnup reasons, was the JM0001 SLOWPOKE-2 in Jamaica in 2015 (see Annex). The MNSR-IAE HEU to LEU conversion was completed in 2016; this was followed by the conversion of GHARR-1 in Ghana in 2017 and NIRR-1 in Nigeria in 2018 (Table 6). The JM0001, GHARR-1 and NIRR-1 conversions were performed with IAEA participation and funded under the United States Global Threat Reduction Initiative (since 2015, the Material Management and Minimization programme), also encompassing the Reduced Enrichment for Research and Test Reactors programme.

76

TABLE 6. LEU CONVERSION SUMMARY (UP TO 2024)

Reactor	Year of LEU conversion	Operating years with HEU
CA0009 SLOWPOKE-2	1997	21
JM0001 SLOWPOKE-2	2015	31
MNSR-IAE	2016	31
GHARR-1	2017	23
NIRR-1	2018	14

There were three main technical differences between the SLOWPOKE and MNSR conversions:

— LEU enrichment was around 19.75% for SLOWPOKE reactors compared with 12.5–13% for MNSRs.
— The LEU approach to criticality methods were different. The fuel cage was fully loaded in the MNSRs during the first approach, but not for SLOWPOKE reactors (see Section 3.11). This difference was due to the use of the Zero Power Test Facility critical facility to establish core-specific loading of the MNSR core pins prior to shipping from the CIAE.
— The MNSR HEU cores were unloaded into a temporary shielding cask (transfer cask), which was located on the top of the reactor container. For SLOWPOKE reactors, the temporary shielding flask was located at the bottom of the pool.

5.2. HEU TO LEU CONVERSION PROJECT ACTIVITIES

Numerous activities, not all involving physical work, are required for a conversion project, and these are summarized in Sections 5.2.1–5.2.4.

5.2.1. Personnel requirements

Significant numbers of personnel are needed from a range of collaborators, including fuel repatriation organizations from China or the USA, the IAEA, the national regulator, specialist contractors and facility staff. A project manager with an authorized budget, who can be available on a continual as-required basis for a number of years during planning and execution, is essential.

5.2.2. LEU fuel assembly fabrication

Fresh LEU fuel fabrication involves the following generic steps:

— Acquisition of LEU (of required enrichment) oxide;
— Testing of UO_2 powder sinterability;
— Pellet fabrication;
— Preparation of Zircaloy-4 cladding, end caps and fuel cage components;
— Fabrication of fuel elements;
— Fuel cage assembly.

5.2.3. HEU repatriation

SLOWPOKE HEU, which was originally supplied by the USA, has been repatriated to the United States Department of Energy Savannah River site. AECL used a dedicated shipping cask (F-257) for the HEU from Canadian facilities. JM0001 used the US NAC-LWT shipping flask (see Annex) [82]. MNSR HEU fuel was repatriated to China, the original supplier. The HEU repatriation from Ghana was supported by China, the USA and the IAEA. The repatriation from Nigeria was supported by China, Norway, the United Kingdom, the USA and the IAEA. The shipping cask used for the MNSR fuel from Ghana and Nigeria was the TUK-145/C-MNSR. Overflight permits in transit countries were also required for the MNSR fuel transfers that were made by road and air.

5.2.4. Safety and licensing

Safety analyses and licensing are required for all main HEU to LEU conversion project activities. The most important safety and licensing activities include:

— Approvals to shut down the reactor for conversion and to remove the reactor core;
— Approvals for site modifications;
— Approvals to import and export dual use equipment;
— Approvals of transport packages;
— Licences to import LEU and export HEU fuel;
— Licences to handle, use and transport nuclear material;
— Licence for reactor operation with LEU.

5.3. LEU TO LEU REFUELLING

The RMC SLOWPOKE was the first of the reactors to have an initial fuel charge of LEU in 1985. After 36 years of operation, the initial RMC LEU fuel was replaced in 2021 (see Annex). The only other reactor to have had an LEU first fuel charge was MNSR IHNI-1. The SUT MNSR also will have an LEU first fuel charge. The University of Toronto SLOWPOKE-1 and the Kanata Isotope Production Facility SLOWPOKE-2 reactors (see Table 2) were also refuelled, but with HEU, after both reactors were transferred from their previous locations (see Table 2).

5.3.1. Fresh fuel supply and spent fuel return

For fresh LEU fuel, potential future issues for both SLOWPOKE reactors and MNSRs could be fuel manufacturing capability and the cost and availability of enriched fuel. The fuel element design itself is not complex, consisting of only four basic components and four components for the fuel cage structure. The elements are relatively easy to manufacture, compared with some other research reactor fuels. A number of countries currently have the capability of producing CANDU fuel, which forms the basis for the fuel design, but with dimensional differences and element end plug differences. This manufacturing capability would be expected to be available in the longer term for refuelling. The fuel cage structure is unique, but simple to manufacture for both SLOWPOKE reactors and MNSRs, with only some dimensional differences between them.

Future LEU fuel cost over decades cannot be predicted owing to uncertainties in current LEU enrichment capability. Fuel financing capability will also vary substantially between facilities in different countries. With long lifetime cores around 20–30 years, this implies a large upfront refuelling capital cost, which would also include a one-time infrequent tooling manufacturing cost. The impact of enrichment capability and the upfront fuel manufacturing costs is something that cannot be predicted beyond the next decade for existing SLOWPOKE and MNSR facilities and potential new facilities.

As outlined in Section 5.2.3, there have been no problems with spent fuel repatriation to the USA and China, the sources of all enriched fuel supplied to date for SLOWPOKE and MNSR research reactors, for any of the decommissioned or refuelled reactors. Repatriation arrangements are specified in original reactor purchase agreements for all existing SLOWPOKE reactors and MNSRs [83].

6. HISTORICAL OPERATIONAL EXPERIENCE

This section summarizes the operational experience with SLOWPOKE reactors and MNSRs, including operational reliability, safety record, radiation doses and activity releases.

6.1. OPERATIONAL RELIABILITY AND SAFETY RECORD

More than 50 years of generic operational reliability experience for both SLOWPOKE reactors and MNSRs has not produced any failures resulting in long term outage repairs. The most significant historical deficiency has been that some of the original SLOWPOKE HEU fuel caused increased fission product activity levels in the reactor container water due to manufacturing defects in some fuel element welds [38, 39]. These defects were small enough to not preclude or limit the long term continued operation with the HEU fuel in any of the reactors. For a short time during 1991, the fission product activity level in the reactor container water at the University of Toronto SLOWPOKE resulted in shorter operational periods at 20 kW being required by the regulator.

Historically, airborne activity releases, from purging the HEU reactor container's gas space, have been maintained below regulatory release limits. All three currently operating SLOWPOKE-2 reactors (as of 2024) are LEU fuelled and show significantly lower reactor container water activity levels than the HEU fuelled cores. Coolant activity levels in the RMC SLOWPOKE LEU fuel have been about three orders of magnitude lower than in the HEU reactors [41]. Inspection of individual HEU elements was not feasible, but a remote underwater visual inspection of the Polytechnique Montréal HEU core was made in 1992 by lifting the fuel cage to expose its entire length [38]. Observation of the peripheral 51 elements and some proportion of the inner pins was then possible. No corrosion was identified and no loss of structural integrity was found. Operation with the HEU fuel cage continued until it was subsequently removed during the 1997 LEU conversion, having achieved 21 years of burnup.

The prototype MNSR-IAE experienced some fission product release from an HEU fuel defect in October 1998. This subsequently led to long term fission product activity monitoring of the reactor container water. The increase in reactor water fission product activity has been small enough that operation of the MNSR-IAE reactor has continued to date. The gaseous fission product releases to the environment from the defective fuel have not resulted in the national annual public dose limit being exceeded. No other historical fuel defect events have been reported (as of 2024) in other MNSRs. To date, five SLOWPOKE reactors and MNSRs have performed successful conversion fuelling from HEU to LEU, a feature that was not specifically planned for in the original design.

The highest historical annual average achieved for energy production for all SLOWPOKE reactors was the University of Toronto SLOWPOKE-2 with 13 200 kW·h per year for an operating lifetime of 22.3 years (see Annex).

A summary review of the operational histories of all SLOWPOKE reactors from 1977 to 2014 is provided in Ref. [84], using data from annual reports (except for JM001, for which technical reports were used until 2021). This review, together with additional subsequent operational experience data collected to 2021, showed that the original design and operation related safety objectives have been consistently maintained for all the facilities. To date, no incidents of significant safety concern[28] have been included in any of the annual reports on record since 1995. The operational experience review provided data on reportable events to the Canadian regulator under the SLOWPOKE licence conditions and on events that were not significant enough to be classified as reportable. Generally, reportable events would range from those having a potential safety concern to those with a significant actual safety concern. Non-reportable but documented events typically represent those related to production reliability, mainly reflecting maintenance, testing and inspection issues. Most of the non-reportable types of event would be representative of anticipated operational occurrences, as defined in Ref. [85].

The prototype MNSR-IAE has been operating since 1984 with no significant safety related incidents and only one reportable event. The beryllium reflector, reactor container, control rod and connected major in-core components have not required any replacement, nor any significant maintenance. The NAA irradiation system, reactor control system and reactor water purification system have all been improved. The neutron flux measuring fission chamber was replaced. The defective HEU fuel core was inspected in 1998. After this inspection, the HEU core was approved by the NNSA for continued operation until the 2016 LEU conversion. In 2018, the electrical components used for core temperature monitoring became unreliable and were replaced. In 2015 and 2021, within the scope of IAEA missions, underwater cameras were used to inspect the corrosion condition of the reactor container, which was found to be satisfactory for continued long term operation. As an example of the longest lifetime overall energy production for MNSR HEU cores, the MNSR-IAE operated 31 years for

---

[28] These are defined as incidents resulting in any reactor staff or the public receiving a dose above regulatory limits, any significant release of activity, power excursions or loss of coolant events leading to fuel damage, or significant reactor equipment damage requiring lengthy repair.

a total of 195 000 kW·h, averaging about 6300 kW·h per year. The HEU lifetime annual average number of irradiated samples was about 5000 per year.

Since 1977 there have been eight SLOWPOKE events reportable to the regulator.[29] The first two reportable events, both losses of regulation (1977 and 1981), are discussed in Ref. [63]. The cause of the 1977 event was a flux detector zero signal that resulted in full withdrawal of the control rod while the reactor was in remotely attended operation, with no operators in attendance overnight. A reactor power transient occurred as expected, but otherwise there was no impact or fuel damage. A subsequent control system design change in all SLOWPOKE reactors eliminated this particular fault. The 1981 event was a loss of an analogue control signal from a failed transistor. As a result of the 1977 design change, the reactor was shut down by the control system without any power transient in the 1981 event. Two subsequent reportable events, in 1991 and 2002, also with no impact or fuel damage, were control system problems caused by identical faults in the flux detector switch on the analogue control system consoles [86]. A fifth reportable event, in 1997, was high radiation field (20 kW) alarms below the pool concrete shielding caused by a failed pool water level switch [87]. The sixth reportable event, in 2002, was personnel contamination from an irradiated sample [88]. A seventh reportable event, in 2011, with no impact or fuel damage, was a power transient due to dropped beryllium shims and removal of the control rod during decommissioning [89]. The eighth reportable event, in 2016, occurred when a control rod was stuck in the fully out position, exceeding the operational limit allowed for the allowable time for a flux $>1.4 \times 10^{12}$ cm^{-2}·s^{-1} [90].

From the SLOWPOKE operational experience review of 1976–2021, certain event types not significant enough to be reportable were evident and are relevant for both SLOWPOKE reactors and MNSRs. This summary review of SLOWPOKE anticipated operational occurrences showed 100 events distributed among the following categories:

— Irradiation facilities: 43%;
— Analogue control system and control rod: 31%;
— Auxiliary systems: 22%;
— Shim operations: 4%.

Events in the category with the highest frequency showed the importance of maintaining the integrity of the plastic transfer tubing and the internal condition[30] of the in-core irradiation tubes by appropriate replacement and inspection

---

[29] For JM0001, there have been no events to date that would have been reportable under Canadian licence conditions.

[30] Moisture collection as well as physical inspection.

frequencies, respectively. With regard to auxiliary systems, out-of-reactor water leakage events seemed to dominate; these are likely due to the original use of non-metallic tubing for some of these systems as well as to the lack of redundancy in the original design for various components. High level automatic make-up pool water addition events, pool water overflow line back flows and low pool levels have also occurred.

For the prototype MNSR-IAE, the one historical reportable event was the HEU fuel defect in October 1998, as noted above [91]. For non-reportable events, the most significant has been the replacement in MNSR-IAE of some aluminium alloy irradiation tubes owing to corrosion in 1998.

6.2. SLOWPOKE AND MNSR PERSONNEL RADIATION DOSES

Typical radiation fields in all SLOWPOKE and MNSR facility reactor rooms (excluding transient radiation fields when performing sample irradiations and localized contact radiation fields on purification system components) at full power have all been reported at <50 µSv/h, and typically far below that value. This level is currently being met for all operating LEU SLOWPOKE reactors and HEU and LEU MNSRs.

As noted in Section 3.1.2, the Canadian SLOWPOKE-2 HEU reactors required concrete shielding slabs over the pool owing to fission product releases from the HEU fuel into the reactor container water when the reactors were at power. The reactor shutdown alarm levels for full power operation were set at 0.1 mSv/h. The full power radiation fields above the reactor pool at the HEU JM0001 SLOWPOKE had historically been acceptably low (<50 µSv/h) so that no concrete shielding slabs over the pool were required at JM0001.[31] The RMC reactor, with its unique vertically oriented beam tube (see Annex), produces higher radiation levels on the I-beam support above the pool water level when operated at full reactor power for radiography and uses a high level alarm of 200 µSv/h, which is higher than in other SLOWPOKE reactors [92]. Similarly, the large vertical neutron radiography beam tube in the ENTC MNSR has higher fields, with 250 µSv/h at full power operation being reported. There are no credible accidents postulated during reactor operation in which unanticipated reactor room radiation dose rates would be high enough to require rapid evacuation of the reactor room. None of the MNSR facilities have required concrete radiation shielding covers over the pool.

---

[31] With LEU since 2016, the field at the same location in JM0001 has been about 40 µSv/h at full power.

Radiation fields around the reactor water purification system ion exchange and filter components, typically located in the reactor room in SLOWPOKE reactors and in the reactor building in MNSRs but shielded by concrete and lead, have the highest external accessible radiation fields in all facilities, as expected. The historical reported maximum field on contact with the reactor water purification system ion exchangers was ≈3 mSv/h in 1992 for the University of Toronto SLOWPOKE. These fields were due to the long lifetime fission products from fuel defects, mentioned in Section 6.1, which built up slowly over time as burnup progressed. This reactor had the highest burnup and energy production of all the HEU SLOWPOKE reactors. This radiation field buildup versus burnup was analysed up to 1995 for Canadian HEU SLOWPOKE reactors in Ref. [38].

High radiation fields could be indicative of fuel defects or tramp uranium on the fuel element cladding. Use of SLOWPOKE LEU fuel has resulted in significantly smaller fields compared with HEU fuel. No fission product activity has ever been reported in the pool waters of any of the SLOWPOKE reactors or MNSRs. SLOWPOKE facility personnel radiation doses have always been sufficiently low in all the facilities — low enough that no Canadian facility staff or authorized user had to be mandated as a nuclear energy worker, or NEW, by the regulatory body. All categories of staff have historically received less than the national annual public whole body dose of 1 mSv. One extremity dose of 1.5 mSv from an irradiated sample was noted in Section 6.1 as a reportable event. This SLOWPOKE low operating dose history, documented in the annual reports submitted to the regulator, provides satisfactory demonstration of achieving the radiation protection design safety objectives for NEW and other authorized staff (see Section 3.10.1).

An example of lower personnel radiation doses after LEU conversion is available from prototype MNSR-IAE data. Before the 2016 LEU core conversion, the average annual dose was 0.6 mSv. Since the core conversion, the average annual dose to date has been 0.5 mSv. Since MNSR-IAE was put into operation, the maximum recorded annual personnel dose has been 2 mSv in 1984 [93].

6.3. SLOWPOKE AND MNSR PUBLIC DOSES AND ACTIVITY RELEASE

With regard to public doses from SLOWPOKE reactors and MNSRs, specific dose data for the off-site public are not measured and, as is the case for research and power reactors internationally, the values are inferred via active gaseous and liquid releases. Detailed annual dose calculations for the public, using measured gas sample analysis of gaseous releases from the SRC SLOWPOKE,[32] provide useful benchmark data. The SRC public dose calculation method would be applicable for all SLOWPOKE and MNSR facility operations, recognizing that the operational and location differences would need appropriate modification. Sections 6.3.1 and 6.3.2 provide some public dose information from the gaseous and liquid releases.

6.3.1. Active gaseous releases

There are three main sources of gaseous releases:[33]

— Diffusion of gaseous ^{41}Ar activity out of the in-core irradiation tubes from irradiation of natural ^{40}Ar in the air inside the tubes and the associated return air lines near the core, up from the core location through the irradiation system equipment, past the gaseous exhaust filter to the reactor room ventilation exhaust duct, from where ^{41}Ar is then released during reactor operation at power;
— ^{41}Ar from irradiated sample transfer operations, during reactor operation at power;
— ^{41}Ar from dissolved air in the reactor container water as well as ^{41}Ar, ^{133}Xe, ^{133m}Xe, ^{131m}Xe and ^{135}Xe from the reactor container gas headspace during the weekly gas purge with the reactor shut down (see Section 3.1.3.1).

SRC SLOWPOKE HEU annual reports from 2008 to 2018 calculate a conservative maximum public dose of $\approx$30 µSv/a, of which >95% was attributable to ^{41}Ar. This dose level is well below the annual occupational radiation exposure limits for occupationally exposed workers (50 mSv/a), and is lower than the exposure limits to a member of the general public (1 mSv/a). The fraction of the dose from ^{41}Ar production is independent of fuel type and fuel performance. Regardless of the use of HEU or LEU fuel, historical SLOWPOKE facility experience documented in SRC annual reports indicates that the fission product gas headspace is dominated by ^{133}Xe, with minor concentrations of ^{131m}Xe, ^{133m}Xe and ^{135}Xe and some traces of krypton isotopes. At the time of purging, usually at the beginning of the week following the reactor operation of the previous week, ^{41}Ar will normally have decayed. This reduces potential releases to the environment. A similar procedure is followed at MNSRs.

---

[32] The SRC SLOWPOKE was the only facility to perform a detailed weekly gamma ray spectroscopy of the reactor container gas headspace isotopes, with data summarized in SRC annual reports.

[33] Gaseous releases based on SLOWPOKE experience have been dominated by ^{41}Ar, ^{133}Xe and ^{135}Xe.

6.3.2. Active liquid releases

Routine active liquid discharges for SLOWPOKE and MNSR water systems are not required or performed, so annual reports do not record any active liquid discharges. Historically, all reactors have reported very low pool water activity levels.

Routine water samples taken from the reactor container for analysis are typically returned to the reactor container after analysis. These samples are sometimes released to the reactor room water drainage system after analysis, if the activity was acceptably low.

Any water from water system piping leak or break events in the reactor room would be released into the water drainage system, typically designed with floor trenches and a drainage sump. Water from a pool level overflow event would be released into the pool water overflow line and then into the reactor room drainage sump.

Any active liquid wastes from irradiated sample capsules would be stored and periodically monitored to allow for decay until activity levels are indistinguishable from background. Irradiated liquid samples can then be disposed of by normal waste disposal techniques.

7. AGEING MANAGEMENT

Current safety standards for research reactors require that provisions be made in the design to facilitate ageing management. The IAEA Specific Safety Guide No. SSG-10 (Rev. 1), Ageing Management for Research Reactors [94], states in para. 4.1:

"Ageing management of [structures, systems and components] important to safety should be implemented proactively (i.e. with foresight and anticipation) throughout the lifetime of the research reactor; that is, in design, fabrication and construction, commissioning, operation (including utilization and modification), and decommissioning."

The concept of a modern, integrated ageing management programme as described in Ref. [94] did not exist when the first SLOWPOKE reactors and MNSRs were designed and built. The first operation and maintenance documentation did, however, establish clear requirements and procedures for planned ageing related replacements of most major components inside the reactor container (e.g. beryllium reflectors, control rod, in-core instrumentation and irradiation tubes). Two unplanned exceptions were replacements for the whole reactor container, including the lower core support platform, and a new fuel core. The feasibility of replacing a whole reactor container and contents was inherent in the design, as only the uncomplicated container upper support structure was connected to the pool. All external systems and connections to the container are easily maintained, replaceable or upgradable. These involve mainly non-active non-contaminated components, other than some water purification system equipment. As noted in Table 2, two active SLOWPOKE reactor containers and contents have been successfully removed and transported from their original pools and relocated, within just a few days, in new facility pools in different physical locations. In both these cases, new fuel cores were used, demonstrating the ease of refuelling, which had not originally been foreseen or planned for. Procedures were not originally developed for placing new fuel cores into an active reactor container, but the simple design ensured that this was subsequently feasible for both SLOWPOKE reactors and MNSRs.

7.1. AGEING MANAGEMENT ACTIVITIES

The dominant activities of TRIGA reactor ageing management (see chapter 6 of Ref. [1]) are (i) major component in-service inspections, (ii) moderator and coolant pool water chemical control, (iii) fuel inspections and (iv) control system instrumentation and control evolution. These same ageing management activities are equally applicable to both SLOWPOKE reactors and MNSRs. The sections below summarize the equivalent SLOWPOKE and MNSR historical ageing management activities.

7.1.1. Major component in-service inspections

All the reactors have some form of in-service inspection of reactor containers at varying intervals. Inspections of the outside surface of the reactor container is relatively simple; the inside inspection is more complex owing to more limited accessibility but is still feasible.

Since the first commissioning of SLOWPOKE reactors and MNSRs, very few modifications to and ageing related maintenance of the major structures of the pool, reactor containers and internal components have been necessary, as can be seen in the facility reports included as on-line supplementary files (see Annex). In PARR-2, six irradiation tubes were found to be filled with water owing to pitting corrosion and were replaced in the period 2006–2016 (see Annex). As noted in Section 6.1, some MNSR-IAE irradiation tubes were also replaced in 1988 because of corrosion. Corrosion has also been experienced from water condensation inside dry irradiation tubes. Water condensation or in-leakage, can be detected from unexpected changes in reactivity from control rod positioning, as well as from moisture observed on capsules after irradiation. A commissioning event for Polytechnique Montréal in 1976 also resulted in the need to replace an irradiation tube [84]. This resulted from the initial use of stainless steel irradiation capsules that caused impact damage to the bottom of an irradiation tube. Subsequent irradiations all used polyethylene capsules. None of the decommissioned SLOWPOKE reactors or MNSRs have been permanently shut down as a result of major component corrosion.

Inspection of the beryllium reflectors has typically not been performed. The oldest beryllium reflector still in service is at Polytechnique Montréal. Cumulative fluences in the reflector beryllium are typically well below those at which irradiation damage might be expected. The operating limit for a beryllium reflector derived from long term operational experience in a 10 MW thermal research reactor was 26 000 MW·d [95]. The University of Toronto SLOWPOKE-2 reactor indicates that 22 continuous years of operation required a cumulative energy expenditure of 12 MW·d (see Annex). The predicted ageing limit for SLOWPOKE and MNSR reflector beryllium, under expected operating conditions, is thus well beyond the conceivable lifetime of any reactors of these types.

Of lesser longer term significance than inspections of major components, but of shorter term safety importance, is the need for inspection and/or periodic replacement of the polyethylene capsules routinely used for ASDS testing to prevent embrittlement from radiation and ageing. Irradiation system plastic tubing in the reactor room leading to and from the reactor container irradiation sites also requires ongoing periodic replacement owing to ageing embrittlement, which can lead to cracks and air leakage. The replacement frequency will vary with facility operation, the type of tubing and local environmental conditions. Upgrading of capsule irradiation equipment in the reactor room has occurred in all facilities, as this technology has been developed and upgraded. Such changes typically do not impact reactor safety or the basic reactor design features in a significant way.

7.1.2. Reactor container water and pool water chemical control

Corrosion of the reactor container and all its components is minimized using water chemistry quality control for both the pool water and reactor container water (see Section 3.8). No major historical component corrosion events have occurred in any of the reactors, as noted in Section 7.1.1, other than in some irradiation tubes, which are designed to be easily replaced.

7.1.3. Fuel inspections

Owing to the fuel cage design, only very limited in-service inspection is possible for SLOWPOKE and MNSR fuel (see Section 6.1 and Ref. [38]). Routine fuel inspections are therefore not performed. The onset of any fuel defect can be easily and quickly detected from fission product gas activity in the reactor container gas space or increased activity of the reactor container water. A potential disadvantage of these reactors is that individual defective fuel elements cannot be replaced. However, the low power rating of the fuel limits the likelihood and extent of fuel defects and fission product releases. To date, no fuel assembly has ever had to be removed because of a fuel defect.

7.1.4. Control system instrumentation and control evolution

The original control systems of all reactors have already been, or are planned to be, changed to digital systems, incorporating new and upgraded digital monitoring systems. The latter are typically more comprehensive than those provided in the original analogue control designs because of the significant advances over the decades. The original analogue control systems needed few changes over the decades, with the availability of spare parts eventually becoming the limiting factor, rather than loss of functional capability from component failures. Polytechnique Montréal currently operates the oldest original control system after 47 years, with adequate spares being available from decommissioned SLOWPOKE reactors.

7.2. TECHNICAL SUPPORT AND AVAILABILITY OF REPLACEMENT PARTS

This section summarizes the technical support and availability of replacement parts for SLOWPOKE reactors and MNSRs.

7.2.1. SLOWPOKE reactors

Technical support for all the Canadian SLOWPOKE reactors was initially provided by AECL as part of its Commercial Products Division in Ottawa, starting with the prototype SLOWPOKE-1 when the reactor was transferred from CRL to the University of Toronto in 1971 (see Table 2). From 1971, various nuclear component replacement parts and specialized tooling were available for use from AECL for all SLOWPOKE reactors. In September 1988, the technical support for SLOWPOKEs was transferred from AECL to the newly privatized company

Nordion International, Ottawa.[34] After 2000, the technical support was then transferred back to AECL, CRL, where it remained until AECL was restructured as a federal government owned, contractor operated operation, effective in 2015. Since 2015, Canadian Nuclear Laboratories has been contracted to provide the technical support for SLOWPOKE reactors.

For the Canadian SLOWPOKE-2 reactors, the CNSC requires that two categories of licensed personnel provide technical support: a reactor engineer and a reactor technician. The licensing consists of (i) past and continuing training according to a written programme accepted by the CNSC, (ii) passing written tests and (iii) supervised experience of work inside a SLOWPOKE-2 reactor container. The current regulatory requirements for nuclear maintenance support are that the following support activities can only be conducted by, or under the direct supervision of, the reactor engineer:

— Breaking the reactor container security seal and resealing the reactor container;
— Opening and closing the reactor container;
— Providing nuclear maintenance activities to the reactor;
— Removing or replacing the fuel as authorized by the regulator;
— Modifying the reactor as required and authorized by the regulator;
— Supervising maintenance of the control system;
— Establishing the reactor excess reactivity and calculating the desired reactivity changes.

The reactor technician support activities are:

— To perform any required changes in the reactor container or modifications to the core and reflectors under the direction and supervision of the reactor engineer;
— To diagnose, repair and replace components that are outside the reactor seal, or inside the seal but above the reactor pool.

Thus, in Canada, the licensing and training of technical support personnel are likely to remain with an organization such as the Canadian Nuclear Laboratories but under AECL authority. With only two remaining Canadian reactors, the cost of purchasing and maintaining technical support for a facility based on very low operating costs could conceivably become a longer term issue.

---

[34] In 1991 Nordion International Inc. was sold to MDS Health Group, and the name was changed to Nordion Inc. in 2010.

As the nuclear regulatory body of Jamaica has been in the early stages of development since 2015, JM0001 has continued its historical use of IAEA advice and IAEA technical support, particularly with respect to significant projects (e.g. the LEU conversion programme and a programme initiated in 2020 to upgrade to a digital control system). Prior to 2015, the International Centre for Environmental and Nuclear Sciences, SLOWPOKE was not subject to the CNSC technical support requirements noted above. However, the facility ensured that all work inside the reactor container, mainly beryllium shim additions, was supervised by personnel qualified by the CNSC for similar work in Canada. Since the conversion in 2015, JM0001 staff have adequate experience in performing all the activities of the reactor engineer and reactor technician described above. For the conversion programme (see Annex), JM0001 purchased new and updated specialized tooling from Polytechnic Montréal and are now adequately self-sufficient for anticipated operational and maintenance support activities including the current control system. The recent upgrading of the instrumentation and control system involved an external supplier.

Currently, neither operational SLOWPOKE has an additional beryllium top reflector (i.e. a big shim or doughnut; see Section 2.2.3) installed. The JM0001 SLOWPOKE now has the University of Toronto's big shim in storage, for possible future use.

As all the SLOWPOKE reactors aged, and as defects or failures occurred, original replacement components became more difficult to acquire. Spare parts have been made available to currently operating SLOWPOKE-2 reactors as other reactors were decommissioned (see Table 2). The majority of replaced components were from the various auxiliary systems (e.g. water purification systems). Such components have been relatively easy to update and have been replaced over the years through commercial suppliers, as meeting the rigorous standards for replacement of nuclear grade components of power reactors was not necessary. Most of the JM0001 specialized tooling needed for the LEU conversion was made by Polytechnique Montréal. The remaining SLOWPOKE-2 reactors have upgraded original auxiliary system designs to meet appropriate electrical, instrumentation and building codes.

For the RMC SLOWPOKE-2, the original Mark 2 analogue control system was the most significant component replacement. The JM0001 SLOWPOKE digital control system installation was completed in 2023. Typically, as instrumentation has developed over the decades, all facilities have improved process parameter monitoring and alarm capabilities since the original design. The Polytechnique Montréal SLOWPOKE is anticipating replacing their original analogue control system with a digital system in the future. Currently, some spare parts are available from other shutdown facilities. The top level overriding requirements for the DCC will be the use of a supplier and/or designer with established experience in research reactor DCC systems and an independent review of the DCC design and operation confirming that the original control system design requirements can be met, or exceeded, with the analysis rationale documented.

SLOWPOKE LEU fuel has provided adequate burnup and failure-free performance for 37 years. If a requirement for refuelling arises, there may be difficulty finding a supplier that is willing to tool up to make such a relatively small amount of fuel for one-time production. With the core lifetime being predictable with a high degree of accuracy, a facility would have many years available for the planning process.

7.2.2. MNSRs

Technical support for the prototype and other MNSRs has been provided by the CIAE. For China's domestic MNSRs, operations personnel need to have an operator certificate issued by the Chinese nuclear security bureau. Prior to certification, a period of continuous training and testing by certified operators is required. For foreign MNSRs, China trains a group of operators who then manage the operations themselves. Chinese operators are divided into operators and senior operators. Some activities can only be carried out under the direct supervision of senior operators, in accordance with regulatory requirements:

— Maintenance work on the top of the reactor;
— Addition of top beryllium reflectors;
— Replacement of the reactor core;
— Maintenance activities on the control system.

The main tasks of the operator are:

— Routine reactor startup, operation and shutdown;
— Planned work under the direction and supervision of senior operators;
— Repair and maintenance of reactor auxiliary systems.

Normal operation of the MNSR requires a minimum staffing level of only one operator. Repairs and maintenance usually require only a small number of personnel, so the operational cost is quite low. The MNSR owner also can engage operators from the prototype MNSR laboratory. Technical staff for operation and maintenance can be provided by the CIAE through a technical service contract. This is the case, for instance, for the operators and maintenance personnel of the MNSR IHNI-1.

After decades of operation, the MNSRs are ageing, and some components need to be replaced. In China, the MNSR components are provided by the CIAE. Components not affecting reactor safety can be directly purchased commercially and replaced. For overseas MNSRs, some components were supplied for the facilities by the CIAE. The facilities can then replace these components. In the event that external technical support and equipment are required, CIAE provides support for equipment and qualified maintenance staff. The Ghanaian and Nigerian MNSRs have both received technical support from the CIAE.

The fuel core of the MNSR is an integral component that, based on the operational experience of the prototype MNSR and other reactors, can provide enough burnup to operate safely for more than 30 years. MNSR LEU conversions have been supported by the CIAE, which has to date completed conversions of the prototype MNSR-IAE, GHARR-1 and NIRR-1 reactors. Conversions are expected to be completed for all other MNSRs. The MNSR LEU fuel has to date provided adequate burnup and failure-free performance for 36 years.

As discussed for SLOWPOKE reactors, finding a supplier that is willing to produce such a relatively small amount of fuel may be difficult. However, with MNSR LEU fuel being almost completely interchangeable, the larger combined potential fuel requirement would be expected to secure at least one supplier.

8. DECOMMISSIONING

This section summarizes the current decommissioning status of SLOWPOKE reactors and MNSRs and the planning and preparation required for decommissioning.

8.1. CURRENT DECOMMISSIONING STATUS

Decommissioning is the last stage in the life cycle of nuclear facilities and "refers to the administrative and technical actions taken to allow the removal of some or all of the regulatory controls from a facility" [96]. Typical technical activities performed during decommissioning involve decontamination, dismantling, demolition, management of waste, and removal of any remaining structures and soil if contaminated, leading to the release of the site from regulatory control. Authorization of the decommissioning actions, depending on the country, includes the approval by the national regulatory body of the final decommissioning plan and supporting documents.

Requirement 6 of IAEA Safety Standards Series No. GSR Part 6, Decommissioning of Facilities, states that licensees are responsible for performing "environmental impact assessments in support of decommissioning actions" [96]. However, each Member State implements its own regulations and procedures. The environmental impact assessment assesses the risk of decommissioning activities with environmental significance and specifies how the compliance with environmental requirements will be ensured. After the completion of decommissioning, the regulatory body reviews and approves the final decommissioning report, submitted by the licensee. The site is typically then released from regulatory control for free unrestricted use or may remain with some restricted site usage conditions.

The main decommissioning activities for SLOWPOKE reactors and MNSRs are:

— Defuelling preparation;
— Fuel removal;
— Reactor container and core component removal;
— Removal of auxiliary systems;
— Pool cleanup;
— Removal and transport of waste to final destination(s);
— Final radiological survey;
— Civil works restoration.

Each of these activities comprises detailed requirements for personnel, equipment and tooling, safety assessments, radioactive waste characterization and overall waste management procedures.

As of 2024, seven SLOWPOKE reactors and two MNSRs had been decommissioned (see Table 2). For all these reactors, fuel has been repatriated and all reactor related equipment, active components and materials have been removed from the reactor facility room or building, achieving unrestricted access and utilization for the rooms occupied by the reactor facility. Waste storage facilities at AECL CRL were used for SLOWPOKE active waste. MNSR waste was sent to north-western Gansu Province's low and intermediate level waste repository site, leaving all facility sites available for future unrestricted use. With this established experience, no significant problems are predicted for the time when eventually all MNSRs and SLOWPOKEs shut down and enter their decommissioning phase. All other States with MNSRs, as well as Jamaica with its SLOWPOKE reactor, will need to plan in advance the location for active waste storage, other than for fuel for which repatriation was established at the time it was procured.

The initial reactor generic decommissioning plan procedures for SLOWPOKE reactors were part of the original safety documentation in 1991 [97]. The last four SLOWPOKE reactor decommissioning and facility abandonments (the same as free release of the site, in IAEA terminology) were the University of Toronto (2001), Dalhousie University (2011), University of Alberta (2018) and SRC (2021) (see Table 2), and these were authorized in Canada based on a three stage licensing process for future unrestricted use of the facilities:

— An environmental assessment screening;
— A licence to decommission;
— A licence to abandon the facility.

Records since 2006 of the CNSC regulatory proceedings for each of these four decommissioning licensing processes, including reasons for decisions, are publicly available on the CNSC web site. Prior to 2001, for the other shutdown SLOWPOKE reactors, the first stage was not needed but licences to decommission and abandon were required for the buildings, except for the CRL prototype, where the building and reactor room is still in use by AECL. For the 2001 University of Toronto SLOWPOKE-2, decommissioning the concrete and rebar below the reactor satisfied the IAEA unconditional free release limits [98]. For the 2011 Dalhousie SLOWPOKE-2, decommissioning some concrete and rebar specific activities below the reactor were higher than the unconditional free release criteria of the IAEA [99]. A decommissioning plan was therefore developed to remove the affected concrete for disposal as solid radioactive waste. The results of the calculations performed for the Dalhousie decommissioning to predict accurate radioisotope concentrations in the pool structural concrete and rebar can provide useful information for the waste management aspects of other facilities' final licence to abandon, to minimize the amount of active concrete removal required.

The MNSR-SH, which was shut down in 2006, was the first MNSR to be decommissioned. Decommissioning was started in 2008. The reactor core was removed from the reactor container and transported off-site in the same year. Residual active waste was removed from the site to the disposal site. The MNSR-SD was decommissioned in 2008 [100]. The decommissioning scope included all reactor systems, equipment and buildings, and preparation for decommissioning. The main active components were the reactor container, reactor pool, auxiliary systems, ventilation system and some operating waste. All the waste was transported to the disposal site. The reactor site residual activity measurement showed that the site satisfied the requirement for future unrestricted use.

The supplementary files (see Annex) also provide some information on the main licensing decommissioning actions, noting that both the MNSR-SD and MNSR-SH facility locations were authorized for free release. The same three stage decommissioning process is currently under way in China for other nuclear facilities.

8.2. DECOMMISSIONING PLAN AND PREPARATION FOR DECOMMISSIONING

Preparation for decommissioning begins at the design stage and continues throughout the lifetime of the facility. The on-line supplementary files contain information on the status of the decommissioning planning for the facilities that are still in operation (see Annex).

At the current time, there does not appear to be any intention to decommission additional facilities in Table 1 in the near term. To meet current IAEA safety standards, decommissioning draft plans are typically now also documented and periodically updated by the facilities. For the eventual decommissioning of the JM0001 reactor, the operator will provide a revision of the generic decommissioning procedures and of the current plans [101] to support a licence from the Jamaican regulator to eventually decommission the reactor facility and no longer be under regulatory control.

The CNSC states:

"A financial guarantee is a tangible commitment by a licensee that there will be sufficient resources to safely terminate licensed activities. When licensees terminate their activities, they must properly account for the safe disposal of all licensed material and equipment, and must demonstrate that all locations associated with the licence are free of radioactive contamination." [102]

Similarly, in China and for MNSRs in other Member States, the national regulator is responsible for ensuring compliance with decommissioning standards in a way that would satisfy both IAEA guidance and national standards.

9. FUTURE OPPORTUNITIES

Historically, SLOWPOKE reactors and MNSRs have primarily been used for NAA and education and training, and this will likely continue for the foreseeable future. While some reactors have now been decommissioned, the remaining reactors have either relatively recently been converted from HEU to LEU, have been refuelled or are actively planned to be refuelled with LEU. This is an indication that the remaining reactors will continue their contribution to the peaceful uses of nuclear energy for several more years. Owing to their relatively simple design and inherent safety features, SLOWPOKE reactors and MNSRs still provide unique opportunities for nuclear education training and academic research. Their very stable and easily reproducible neutron flux over the reactor lifetime make them ideal neutron sources for activation analysis. The absence of some of the challenges relating to fuel availability, complexity and staffing costs faced by larger research reactors also puts SLOWPOKE reactors and MNSRs at an advantage for their continued operation, provided that their associated systems are upgraded and modernized accordingly.

9.1. NEW BUILDS

There are no major obstacles to building new SLOWPOKE reactors. However, the last conventional SLOWPOKE reactor built was commissioned in 1985 and the SDR was commissioned in 1987, and no plans for new builds have been announced since.

The SUT MNSR is currently the only MNSR planned or under construction. There are no major obstacles to building further new MNSRs.

9.2. UPGRADES AND UTILIZATION PLANS

Of all the SLOWPOKE reactors and MNSRs, only RMC was originally commissioned with LEU fuel. The new SUT MNSR will be commissioned with LEU fuel. Five HEU reactors have been converted to LEU, with current plans to convert the remainder to LEU. Past refuelling and conversion activities have also been the catalyst for several major upgrades of various systems and structures at facilities. Nonetheless, there is still work to be done to continually upgrade and modernize other components and systems. Efforts in the respective ageing management and modernization programmes to facilitate the sustainable and effective operation and utilization of the reactors need to be focused on:

— Digital instrumentation and control systems, compliant with national and international standards and requirements;
— Upgrading of existing and installation of new irradiation facilities;

— NAA automation to increase output, especially since some facilities have a small staff complement;
— Monitoring systems to enable remote monitoring of operating parameters;
— Water purification and cooling systems;
— Reactor mock-up and/or working model to facilitate specialized training for nuclear maintenance (e.g. shim adjustments);
— Implementation of an integrated management system.

There are also safety related upgrades and other activities that need to be prioritized, including:

— Reactor safety documentation (e.g. SARs, OLCs, radiation protection, emergency preparedness and response plan, initial decommissioning plan).
— Radiation monitoring systems, including fixed and portable radiation detectors.
— Emergency power supply, emergency lighting and fire detection systems.
— Periodic visual inspections of reactor pool wall and lining and reactor container internals. Fuel inspection remains challenging since the core is completely surrounded by the reflectors.[35]
— Procurement of replacement neutron detectors and related components when the original components are no longer available.
— Some recent MNSR design modifications have included horizontal and vertical neutron beam tubes and neutron beam accessibility below the reactor core. These are major changes to the original SLOWPOKE and MNSR designs, which required additional analysis of the safety implications.

9.3. FUTURE APPLICATIONS

The main applications of SLOWPOKE reactors and MNSRs are detailed in Section 4. They are generally the same for all the reactors, with only a few reactors having additional capabilities for other applications. The main applications across all of these reactors are NAA, education and training, and isotope production (see Table 5). There are possibilities, however, that some upgrades already in place at some SLOWPOKE and MNSR facilities, such as cadmium-lined irradiation sites for epithermal NAA, could be fairly easily adopted by other facilities to expand the scope of utilization and applications. Other upgrades at some sites, such as

---

[35] Lifting an irradiated fuel cage above the reflectors for inspection has been performed once (see Section 6.1), but this provides only a limited inspection capability due to the compact design of the fuel cage elements.

a neutron beam tube for neutron radiography, cannot be as easily retrofitted. However, there is a potential for SLOWPOKE reactors and MNSRs to be applied in areas such as:

— Materials science and technology;
— Instrumentation development and testing;
— In-pool gamma irradiations (e.g. medical equipment, biological tissue);
— Specialized education and training;
— Nuclear and radiological safety;
— BNCT, radiochemical NAA, PGA, DNC and cyclic NAA;
— Neutron imaging.

The original design of SLOWPOKE reactors and MNSRs had nominal power ratings of 20 kW and 30 kW, respectively, which limited some applications due to their relatively low thermal flux. On LEU conversion, some MNSRs increased the power to 34 kW but still used the same lattice design as the HEU core. However, detailed neutronic and thermohydraulic analyses on SLOWPOKE reactors have indicated that, with modifications, the neutron flux level within the irradiation sites could be increased by ≈75% with only a ≈50% increase in reactor power, while not adversely affecting reactor safety [28]. This increase in neutron flux could potentially broaden the scope of applications of SLOWPOKE reactors and MNSRs, or at the very least, offer new opportunities for expanding existing applications. The SUT MNSR is planned to have 45 kW of power in order to obtain an adequate neutron flux for BNCT.

With the majority of the reactors now being in developing countries, the reactors and their staff will continue to play a vital role in contributing to solving challenges related to development. Operating staff at these facilities are experts in a variety of nuclear related fields, and it will continue to be important that they make available their expertise both nationally to public bodies including government ministries, regulatory bodies, hospitals, universities and industry, as well as to international organizations such as the IAEA.

REFERENCES

[1]   INTERNATIONAL ATOMIC ENERGY AGENCY, History, Development and Future of TRIGA Research Reactors, Technical Reports Series No. 482, IAEA, Vienna (2016).
[2]   DYSON, F.J., Disturbing the Universe, Basic Books, New York (1979) 94–103.

[3]   JARVIS, G.A., MILLS, C.B., Critical Mass Reduction, LA-3651, Los Alamos Scientific Laboratory, Los Alamos, CA (1966).

[4]   AMERICAN NUCLEAR SOCIETY, Nucl. News **10** (1967) 13.

[5]   AMERICAN INSTITUTE OF PHYSICS, LASL builds the world's smallest nuclear reactor, Phys. Today **20** (1967) 68.

[6]   HILBORN, J.W., LYON, R.B., Preliminary Proposal for a Low Cost Neutron Source, Atomic Energy of Canada Limited, Chalk River (1967).

[7]   KAY, R.E., STEVENS-GUILLE, P.D., HILBORN, J.W., JERVIS, R.E., SLOWPOKE: A new low-cost laboratory reactor, Int. J. Appl. Radiat. Isot. **24** (1973) 509–518.

[8]   SLATER, I.J., Small Simple and Safe but Very Hard to Sell: A Technical and Institutional History of the SLOWPOKE Nuclear Research Reactor, MA Thesis, Univ. Toronto (1998).

[9]   INTERNATIONAL ATOMIC ENERGY AGENCY, Safety of Research Reactors, IAEA Safety Standards Series No. SSR-3, IAEA, Vienna (2016).

[10]   HILBORN, J.W., STEVENS-GUILLE, P.D., "First critical: The original SLOWPOKE reactor" (Proc. Tech. Mtg on Low-Power Critical Facilities and Small Reactors, Ottawa, 2010), Canadian Nuclear Society, Toronto (2011).

[11]   YONGMAO, Z., "A safe private nuclear tool, the miniature neutron source reactor", Proc. Sixth Pacific Basin Nuclear Conference, Beijing (1987) 253–259.

[12]   SNSR DESIGN GROUP, SNSR Demonstration Reactor Development Progress, Annual Report of China Institute of Atomic Energy, China Institute of Atomic Energy, Beijing (1982).

[13]   INTERNATIONAL ATOMIC ENERGY AGENCY, Technology and Use of Low Power Research Reactors (Report of a Consultants' Meeting, Beijing, 30 April–3 May 1985), IAEA-TECDOC-384, IAEA, Vienna (1986).

[14]   LAN, Y., ZHENG, Z., SHI, Y., ZHENG, W., LI, F., "Prototype MNSR physical start-up and parameter measurements", Annual Report of China Institute of Atomic Energy, China Institute of Atomic Energy, Beijing (1984).

[15]   KEQIANG, R., et al., "SNSR advances in physical computing", Annual Report of China Institute of Atomic Energy, China Institute of Atomic Energy, Beijing (1980).

[16]   KAY, R.E., HILBORN, J.W., POULSEN, N.B., The self limiting power excursion behaviour of the SLOWPOKE reactor results of experiments and qualitative explanation, AECL-4770, Atomic Energy of Canada Limited, Chalk River (1976).

[17]   SLATER, I.J., The SLOWPOKE nuclear research reactor — A brief history, Working Paper 13, Canadian Society for Mechanical Engineering, Ottawa (1988).

[18]   FARADAY, M.A., "Fuel for Canadian research reactors", ANL/RERTR/TM-1 (Proc. Int. Mtg on Reduced Enrichment for Research and Test Reactors Argonne, IL, 1978), Argonne National Laboratory, Argonne, IL (1978).

[19]   WISE, M.E., KAY, R.E., Description and Safety Analysis for the SLOWPOKE-2 Reactor, AECL Commercial Products, CPR-26, Rev. 1, Atomic Energy of Canada Limited, Chalk River (1983).

[20]   TOWNES, B.M., HILBORN, J.W., "The SLOWPOKE-2 reactor with LEU oxide fuel", Canadian Nuclear Society (Proc. 6th Ann. Conf. of the Canadian Nuclear Society, Ottawa, 1985), also in AECL-8840 (1985).

[21] ATFIELD, M.D., "The SLOWPOKE reactor with low enrichment uranium oxide fuel", (Proc. Int. Mtg on Reduced Enrichment for Research and Test Reactors, Gatlinburg, TN, 1986), Argonne National Laboratory, Argonne, IL (1986) 370–379.

[22] SMITH, A.D., TOWNES, B.M., Description and Safety Analysis for the Slowpoke-2 Reactor with LEU Oxide fuel, AECL Radiochemical Company, CPR-77, Atomic Energy of Canada Limited, Chalk River (1985).

[23] MATOS, J.E., LELL, R.M., "Feasibility study of potential LEU fuels for a generic MNSR reactor" (Proc. 2005 Int. Mtg on Reduced Enrichment for Research and Test Reactors, Boston, MA, 2005), Argonne National Laboratory, Argonne, IL (2005).

[24] ATOMIC ENERGY OF CANADA LIMITED, SLOWPOKE-2 Nuclear Reactor, Commercial Products Number K-900, AECL, Chalk River (1980).

[25] PUIG, F., DENNIS, H., GRANT, C.N., PRESTON, J.A., JM–0001 SLOWPOKE–II Research Reactor Supporting Analysis Report for the Conversion from HEU to LEU, ANL/RTR/TM-17/4, Argonne National Laboratory, Argonne, IL (2017).

[26] HILBORN, J.W., KAY, R.E., STEVENS-GUILLE, P.D., JERVIS, R.E., SLOWPOKE at the University of Toronto: A Laboratory Reactor for Neutron Irradiation, AECL-4212, Atomic Energy of Canada Limted, Chalk River (1972).

[27] INTERNATIONAL ATOMIC ENERGY AGENCY, Analyses Supporting Conversion of Research Reactors from High Enriched Uranium Fuel to Low Enriched Uranium Fuel, IAEA-TECDOC-1844, IAEA, Vienna (2018).

[28] OLIVA, A.F., An Analytical and Experimental Study of Direct Container Cooling in the SLOWPOKE-2 Nuclear Reactor, MSc Thesis, Univ. Toronto (1981).

[29] DUNLOP, D., A Canadian hybrid submarine design: A case for the SLOWPOKE-2 Reactor, Can. Nav. Rev. (2020).

[30] LI, Y., et al., "Design and construct of in-hospital neutron irradiator" (Proc. Int. Conf. on Research Reactors: Safe Management and Effective Utilization, IAEA, Rabat, Morocco, 2011), IAEA, Vienna (2011) CD-ROM.

[31] ZHANG, Z., LIU, T., A review of the development of in-hospital neutron irradiator-1 and boron neutron capture therapy clinical research on malignant melanoma, Ther. Radiol. Oncol. **2** (2018) 49.

[32] MINISTRY OF ECOLOGY AND ENVIRONMENT, Annual Report for Nuclear Safety, http://english.mee.gov.cn/Resources/Reports/Annual_Report_for_Nuclear_Safety/

[33] BENNETT, L.G.I., LEWIS, W.J., HUNGLER, P.C., The development of neutron radiography and tomography on a SLOWPOKE-2 reactor, Phys. Procedia **43** (2013) 21–33.

[34] BONIN, H.W., HILBORN, J.W., CARLIN, G.E., GAGNON, R., BUSATTA, P., "Homogeneous SLOWPOKE reactors for replacing SLOWPOKE-2 research reactors and the production of radioisotopes" (Proc. 3rd Int. Tech. Mtg on Small Reactors, Ottawa, 2014), Canadian Nuclear Society, Toronto (2014) 22535764, 13 pp.

[35] SHANNON, C., Conceptual Design of an Organic-Cooled Small Nuclear Reactor to Support Energy Demands in Remote Locations in Northern Canada, MSc Thesis, Royal Military College of Canada (2018).

[36] CHOOPAN DASTJERDI, M.H., MOKHTARI, J., TOGHYANI, M., Design, construction, and characterization of a prompt gamma neutron activation analysis (PGNAA) system at Isfahan MNSR, Nucl. Eng. Technol. **55** (2023) 4329–4334.

[37] CHENGZHAN, G., YONGCHUN, G., JIJIN, G., XIANFA, Z., Safety Analysis Report for the Miniature Neutron Source Reactor (MNSR), MNSR-GN-2, China Institute of Atomic Energy, Beijing (1992).

[38] LEWIS, B.J., HARNDEN-GILLIS, A.C., BENNETT, L.G.I., Fission product release from uranium-aluminium alloy fuel in SLOWPOKE-2 reactors, Nucl. Technol. **105** (1994) 366–380.

[39] HARNDEN-GILLIS, A.C., BENNETT, L.G.I., LEWIS, B.J., Experiments and analysis of fission product release in HEU-fuelled SLOWPOKE-2 reactors, Nucl. Instrum. Methods Phys. Res. Sect. A **345** (1994) 520–527.

[40] PUIG, F., DENNIS, H., SLOWPOKE-2 alternative core loading configurations analysis for highly improved reactor performance, Ann. Nucl. Energy **128** (2019) 216–230.

[41] BURBIDGE, G.A., JONES, R.T., TOWNES, B.M., Commissioning of the SLOWPOKE-2 (RMC) Reactor, AECL, RCC/TR-85-004, Rev. 1, Ottowa (1986).

[42] WINFIELD, D.J., Safety Analysis for the École Polytechnique SLOWPOKE-2 Reactor, RC-1598 (Rev. 1), Atomic Energy of Canada Limited, Chalk River (1998).

[43] JONES, R.T., DENNIS, H., Commissioning of the JM0001 Slowpoke Reactor with Low Enriched Uranium Fuel, ICENS-SLOW-COMR-001, International Centre for Environmental and Nuclear Sciences, University of West Indies, Jamaica (2017).

[44] SMITH, A.D., Specification for Low-Enriched Uranium Dioxide Fuel Elements for Use in the SLOWPOKE-2 Reactor, CRNL-2929, Atomic Energy of Canada Limited, Chalk River (1985).

[45] BERGERON, A., Fabrication of Uranium Dioxide Fuel Pellets in Support of a SLOWPOKE-2 Research Reactor HEU to LEU Core Conversion (Proc. 3rd Int. Mtg on Small Reactors, Ottawa, 2014), Canadian Nuclear Society, Toronto (2014).

[46] BURBIDGE, G.A., WISE, M.E., GARVEY, P.M., Commissioning of the SLOWPOKE-2 (École Polytechnique) Reactor, CPR-23, Atomic Energy of Canada Limited, Chalk River (1976).

[47] NOCEIR, S., EL HAJJAJI, O., VARIN, E., ROY, R., ROZON, D., "Diffusion calculations for the SLOWPOKE-2 reactor using DONJON" (Proc. Canadian Nuclear Society Conf., Toronto, 1997), Canadian Nuclear Society, Toronto (1997).

[48] ROZON, D., KAVEH, S., SLOWKIN: A Simplified Model for the Simulation of Transients in a SLOWPOKE-2 Reactor, IGE-219, Rev. 1, École Polytechnique de Montréal, Montreal (1997).

[49] HÉBERT, A., SEKKI, D., CHAMBON, R., A User Guide for DONJON Version 5, IGE-344, École Polytechnique de Montréal, Montreal (2022).

[50] BENNETT, L.G.I., BEELEY, P.A., KENNEDY, G.G., Comparison of the operational characteristics of the HEU and LEU fuelled SLOWPOKE-2 reactors, IAEA-SM-310/50P (Proc. Int. Symp. on Research Reactor Safety, Operations and Modifications 5, 1989), Atomic Energy of Canada Limited, Chalk River (1989).

[51] EDWARDS, G., JONES, R.T., Commissioning of the École Polytechnique LEU SLOWPOKE-2 Research Reactor, FFC-RRP-087, Atomic Energy of Canada Limited, Chalk River (1997).

[52] AHMED, Y.A., MANSIR, I.B., YUSUF, I., BALOGUN, G.I., JONAH, S.A., Effects of core excess reactivity and coolant average temperature on maximum operable time of NIRR-1 miniature neutron source reactor, Nucl. Eng. Des. **241** (2011) 1559–1564.

[53] LIAW, J.R., MATOS, J.E., MNSR flux performance and core lifetime analysis with HEU and LEU fuels (Proc. Int. Mtg on Reduced Enrichment for Research and Test Reactors, Prague, 2007), Argonne National Laboratory, Argonne, IL (2007).

[54] BURBIDGE, G.A., IRISH, J.D., Commissioning of the SLOWPOKE-2 (UWI) Reactor, CPR-73, Atomic Energy of Canada Limited, Chalk River (1984).

[55] ALETA, C.R., Safety Report, SLOWPOKE-2 (UWI) Reactor, University of the West Indies, Kingston, Jamaica (1985).

[56] INTERNATIONAL ATOMIC ENERGY AGENCY, Safety Analysis for Research Reactors, Safety Reports Series No. 55, IAEA, Vienna (2008).

[57] DUNN, F.E., THOMAS, J., LIAW, J., MATOS, J.E., MNSR transient analyses and thermal hydraulic safety margins for HEU and LEU cores using the RELAP5-3D code (Proc. Int. Mtg on Reduced Enrichment for Research and Test Reactors, Prague, 2007), Argonne National Laboratory, Argonne, IL (2007).

[58] JONAH, S.A., IBRAHIM, Y.V., KALIMULLAH, M., MATOS, J.E., Steady State Thermal Hydraulic Operational Parameters and Safety Margins of NIRR-1 with LEU Fuel Using PLTEMP-ANL Code (Proc. European Nuclear Conference (ENC), Manchester, UK, 2012), European Nuclear Society, Brussels (2012).

[59] INTERNATIONAL ATOMIC ENERGY AGENCY, Guidelines for the Review of Research Reactor Safety, IAEA Services Series No. 25 (Rev. 1), IAEA, Vienna (2024).

[60] UNIVERSITY OF TORONTO, SLOWPOKE Annual Report, University of Toronto, Toronto (1972).

[61] UNIVERSITY OF TORONTO, SLOWPOKE Annual Report, University of Toronto, Toronto (1973).

[62] YOUNGER-LEWIS, D.G., BURBIDGE, G.A., SLOWPOKE-2 Nuclear Reactor Operation and Routine Maintenance, CPSR-362, Rev. 2, Atomic Energy of Canada Limited, Chalk River (1984).

[63] HILBORN, J.W., BURBIDGE, G.A., "SLOWPOKE: The First Decade and Beyond", AECL-8252, Atomic Energy of Canada Limited, Chalk River (1983).

[64] ODOI, H.C., et al., "Completion of Ghana's MNSR conversion from HEU to LEU" (Proc. 38th Int. Mtg on Reduced Enrichment for Research and Test Reactors, Chicago, IL, 2017), Argonne National Laboratory, Argonne, IL (2017).

[65] INTERNATIONAL ATOMIC ENERGY AGENCY, Applications of Research Reactors, IAEA Nuclear Energy Series No. NP-T-5.3, IAEA, Vienna (2014).

[66] INTERNATIONAL ATOMIC ENERGY AGENCY, Strategic Planning for Research Reactors, IAEA Nuclear Energy Series No. NG-T-3.16, IAEA, Vienna (2017).

[67] INTERNATIONAL ATOMIC ENERGY AGENCY, Integrated Research Reactor Utilization Review (IRRUR) Guidelines, IAEA Services Series No. 48, IAEA, Vienna (2023).

[68] BURBIDGE, G.A., SLOWPOKE-2 Product Information Manual, AECL Commercial Products, Isotope Products Development Group, Uranium Geochemical Analysis and Assay by Neutron Activation Analysis/Delayed Neutron Counting, Atomic Energy of Canada Limited, Chalk River (1975).

[69] INTERNATIONAL ATOMIC ENERGY AGENCY, Advances in Neutron Activation Analysis of Large Objects with Emphasis on Archaeological Examples, IAEA-TECDOC-1838, IAEA, Vienna (2018).

[70] INTERNATIONAL ATOMIC ENERGY AGENCY, Development of an Integrated Approach to Routine Automation of Neutron Activation Analysis, IAEA-TECDOC-1839, IAEA, Vienna (2018).

[71] GRANT, C.N., ANTOINE, J.M.R., Instrumental neutron activation analysis in forensic science in Jamaica: The case of the Coral Springs beach theft, Forensic Chem. **7** (2018) 88–93.

[72] ANTOINE, J.M.R., GRANT, C.N., WILLIAMS, J.A., HAMILTON, O.O.J., ROBERTS, C.K., A pilot study on the use of neutron activation analysis and multivariate statistics for the provenance of Jamaican *Cannabis sativa* L (Marijuana), Forensic Sci. Int. **335** (2022) 111303.

[73] JONAH, S.A., SADIQ, U., OKUNADE, I.O., FUNTUA, I.I., The use of the k_0-IAEA program in NIRR-1 NAA laboratory, J. Radioanal. Nucl. Chem. **279** (2009) 749–755.

[74] BOULANGER, A., EVANS, D.J.R., RABY, B.F., Uranium analysis by neutron activation delayed neutron counting (Proc. 7th Ann. Symp. of Canadian Mineral Analysts, Thunder Bay, Ontario, 1975) (1975) 61–71.

[75] INTERNATIONAL ATOMIC ENERGY AGENCY, Neutron Activation Analysis Using Short Half-life Radionuclides, IAEA-TECDOC-2055, IAEA, Vienna (2024).

[76] HANCOCK, R.G.V., University of Toronto SLOWPOKE Reactor Facility Annual Compliance Report for the Period 1997 July to 1998 June, University of Toronto, Toronto (1998).

[77] MATOUSKOVA, J., SCHILLINGER, B., SKLENKA, L., New neutron imaging facility NIFFLER at very low power reactor VR-1, J. Imaging **9** (2023) 15.

[78] MOKHTARI, J., CHOOPAN DASTJERDI, M.H.C., Development and characterization of a large thermal neutron beam for neutron radiography at Isfahan MNSR, Nucl. Instrum. Methods Phys. Res., Sect. A **1051** (2023) 168209.

[79] INTERNATIONAL ATOMIC ENERGY AGENCY, Advances in Boron Neutron Capture Therapy, IAEA, Vienna (2023).

[80] ZHANG, Z., LIU, T., A review of the development of in-hospital neutron irradiator-1 and boron neutron capture therapy clinical research on malignant melanoma, Ther. Radiol. Oncol. **2** (2018) 49.

[81] JONAH, S.A., IBRAHIM, Y.V., AKAHO, E.H.K., The determination of reactor neutron spectrum-averaged cross-sections in miniature neutron source reactor facility, Appl. Radiat. Isot. **66** (2008) 1377–1380.

[82] NAC INTERNATIONAL, NAC-LWT, Legal Weight Truck Cask System, LWT-16B NRU/NRX, Damaged Fuel Application Initial Submittal, Non-proprietary Version, Doc. No. 71-9225 (2016),
https://www.nrc.gov/docs/ML1636/ML16364A006.pdf

[83] INTERNATIONAL ATOMIC ENERGY AGENCY, Return of Research Reactor Spent Fuel to the Country of Origin: Requirements for Technical and Administrative Preparations and National Experiences, IAEA-TECDOC-1593, IAEA, Vienna (2009) CD-ROM.

[84] WINFIELD, D.J., Safety Analysis Report for the Jamaica ICENS SLOWPOKE-2 Reactor, ICENS-SLW-SAR-001, CPR-26, Rev. 2, International Centre for Environmental and Nuclear Sciences, University of West Indies, Jamaica (2015).

[85] INTERNATIONAL ATOMIC ENERGY AGENCY, IAEA Nuclear Safety and Security Glossary, 2022 (Interim) Edition, IAEA, Vienna (2022).

[86] UNIVERSITY OF TORONTO, Annual Compliance Report for the University of Toronto SLOWPOKE, University of Toronto, Toronto (1991).

[87] UNIVERSITY OF TORONTO, Annual Compliance Report for the University of Toronto SLOWPOKE, University of Toronto, Toronto (1997).

[88] SASKATCHEWAN RESEARCH COUNCIL, Facility Annual Compliance Report for the Saskatchewan Research Council SLOWPOKE-2 (2002–2003), SRC, Saskatoon (2003).

[89] CANADIAN NUCLEAR SAFETY COMMISSION, Dalhousie University SLOWPOKE-2 Reactor Report on the Unintended Extraction of the Control Rod, e-Doc 4435896, CNSC, Ottawa (2011).

[90] SASKATCHEWAN RESEARCH COUNCIL, Annual Compliance and Operational Report for the SLOWPOKE-2 Facility at Saskatchewan Research Council, SRC, Saskatoon (2016).

[91] GAO, J., ZHANG, X., LI, Y., Prototype Miniature Neutron Source Reactor, Final Safety Analysis Report. Internal Documents of MNSR, China Institute of Atomic Energy, Beijing (1989).

[92] ROYAL MILITARY COLLEGE OF CANADA, Annual Compliance and Operational Report for the SLOWPOKE-2 Facility at RMC, 2012–2013, Kingston (2013).

[93] CHINA INSTITUTE OF ATOMIC ENERGY, Annual Report of Staff Radioactive Dose Detection. Internal Documents of CIAE, China Institute of Atomic Energy, Beijing (1984).

[94] INTERNATIONAL ATOMIC ENERGY AGENCY, Ageing Management for Research Reactors, IAEA Safety Standards Series No. SSG-10 (Rev. 1), IAEA, Vienna (2023).

[95] BOJANOWSKI, C., et al., Impact of MURR LEU Conversion on Beryllium Reflector Lifetime, ANL/RTR/TM-20/1, Argonne National Laboratory, Argonne, IL (2020).

[96] INTERNATIONAL ATOMIC ENERGY AGENCY, Decommissioning of Facilities, IAEA Safety Standards Series No. GSR Part 6, IAEA, Vienna (2014).

[97] BURBIDGE, G.A., SLOWPOKE-2 Nuclear Reactor Generic Decommissioning Procedure, CPR-32, Atomic Energy of Canada Limited, Chalk River (1991).

[98] SMITH, P.G., EVERALL, D., ARIANI, I., TSANG, K., Comparison of measured and calculated concrete and rebar specific activity during decommissioning of the Dalhousie SLOWPOKE-2 reactor, Radiat. Prot. Dosim. **155** (2013) 369–373.

[99] INTERNATIONAL ATOMIC ENERGY AGENCY, Application of the Concept of Clearance, IAEA Safety Standards Series No. GSG-18, IAEA, Vienna (2023).

[100] LI, Y., et al., Jinan Miniature Neutron Source Reactor Decommissioning, Progress Report on Nuclear Science and Technology in China (Vol.3) (Proc. Acad. Ann. Mtg of China Nuclear Society No. 5 2013), China Nuclear Physics Society, Beijing (2013).

[101] INTERNATIONAL CENTRE FOR ENVIRONMENTAL AND NUCLEAR SCIENCES, Decommissioning Plan for the UWI ICENS SLOWPOKE-2 Reactor, Draft Rev. 0, ICENS, University of the West Indies, Jamaica (2015).

[102] GOVERNMENT OF CANADA, Financial Guarantees (2024), https://nuclearsafety.gc.ca/eng/nuclear-substances/licensing-nuclear-substances-and-radiation-devices/licensing-process/financial-guarantees/

Annex

SLOWPOKE AND MNSR FACILITY REPORTS

The following annex presents abstracts of papers on SLOWPOKE and MNSR research reactors, organized by country. The full papers can be consulted at: https://inis.iaea.org/records/t8xmw-tbx15

Paper ID#	Author	Country	Title
1	R.G.V. Hancock, S. Aufreiter	Canada	A Brief History of SLOWPOKE-1 at the University of Toronto
2	R.G.V. Hancock, S. Aufreiter	Canada	A Very Brief History of SLOWPOKE-2 at the University of Toronto
3	L.G.I. Bennett, et al.	Canada	SLOWPOKE-2 Facility at the Royal Military College of Canada
4	J.W. Hilborn, D.J. Winfield	Canada	The SLOWPOKE Demonstration Reactor Project
5	C. Chilian, D.R. Hall, G.G. Kennedy, J. St-Pierre	Canada	SLOWPOKE Facility at Polytechnique Montréal
6	Y. Li, et al.	China	CIAE Miniature Neutron Source Reactor
7	J. Hong, M. Wang, Q. Hao, D. Ouyang, Y. Li	China	Shandong Miniature Neutron Source Reactor
8	D. Peng, Q. Hao, S.Y. Zou, J.H. Zgang, K.J. Li	China	Shanghai Miniature Neutron Source Reactor

Paper ID# (cont.)	Author	Country	Title
9	H. Zhao, Q. Luo, S. Hu, D. Ouyang, K. Li	China	Shenzhen University Miniature Neutron Source Reactor
10	H.C. Odoi, et al.	Ghana	Ghana Research Reactor-1 Facility Report
11	M.H. Choopan Dastjerdi, J. Mokhtari	Islamic Republic of Iran	Isfahan Miniature Neutron Source Reactor
12	H.T. Dennis, C.N. Grant	Jamaica	The Jamaican SLOWPOKE-2 Research Reactor
13	E. Hagberg, C. Grant	Jamaica	Fuel Conversion of the ICENS SLOWPOKE-2 Reactor at the University of the West Indies, Jamaica
14	S.A. Jonah	Nigeria	The Nigeria Research Reactor-1
15	T. Mahmood, M. Iqbal	Pakistan	Pakistan Research Reactor-2
16	I. Othman, M. Boush, A. Sarheel	Syrian Arab Republic	Syrian Research Reactor
17	C. Kobdaj, et al.	Thailand	Suranaree University of Technology Research Reactor

A BRIEF HISTORY OF SLOWPOKE-1 AT THE UNIVERSITY OF TORONTO

R.G.V. HANCOCK
Department of Anthropology,
McMaster University,
Hamilton
Email: rgvhancock@gmail.com

S. AUFREITER
SickKids Research Institute,
The Hospital for Sick Children,
Toronto
Email: susanne.aufreiter@sickkids.ca

Ontario, Canada

Abstract

The SLOWPOKE reactor was installed at the University of Toronto in order to assess how useful it could be for research in a university/hospital environment, and it was very successful. The reactor became a people friendly analytical resource for the Toronto research community. This history includes people, ideas, events and facility descriptions.

A VERY BRIEF HISTORY OF SLOWPOKE-2 AT THE UNIVERSITY OF TORONTO

R.G.V. HANCOCK
Department of Anthropology,
McMaster University,
Hamilton
Email: rgvhancock@gmail.com

S. AUFREITER
SickKids Research Institute,
The Hospital for Sick Children,
Toronto
Email: susanne.aufreiter@sickkids.ca

Ontario, Canada

Abstract

The SLOWPOKE-2 reactor at the University of Toronto worked admirably well for nearly 23 years, until it was deemed redundant. It provided undergraduate students in university departments from Chemical Engineering and Applied Chemistry, through to Physics and Anthropology, with experience of neutron activation analysis and of some of the useful applications of radioactivity in research. Its main function was to support diverse research projects of students and staff from within the university in the faculties of Arts and Science (Archaeology and Anthropology, Botany, Environmental Studies, Geography, Geology, Physics, and Zoology,), Engineering and Medicine, and also served as a resource for researcher colleagues outside the University of Toronto, locally, nationally and internationally.

SLOWPOKE-2 FACILITY AT THE ROYAL MILITARY COLLEGE OF CANADA

L.G.I. BENNETT, K.S. NIELSEN, H.W. BONIN,
B.J. LEWIS, P. SAMULEEV, P. CHAN
Department of Chemistry and Chemical Engineering,
Royal Military College of Canada,
Kingston, Ontario, Canada
Emails: bennett-l@rmc.ca, nielsen-k@rmc.ca, bonin-h@rmc.ca, pavel.
samuleev@rmc.ca, paul.chan@rmc.ca

Abstract

The SLOWPOKE-2 Facility at the Royal Military College of Canada is located in Module 5 of the Sawyer Building on two adjacent floors with a low enriched uranium fuelled reactor in an open pool on the first floor and more laboratory space on the second floor. Suspended in this vertical pool is the reactor container with the reactor core situated near the bottom, surrounded by inner and outer irradiation sites, relative to the beryllium annulus. In the pool are located a neutron beam tube for neutron imaging as well as three elevators for irradiations of large samples. The facility is used for teaching, research and service work in applications of neutron activation analysis, low level counting, delayed neutron counting, radioisotope production, irradiation and neutron imaging. The reactor has been modernized with a digital control system as well as upgraded electrical connections with more upgrades planned for the various irradiation controllers. The reactor has been refuelled in 2021 after 36 years of operation.

THE SLOWPOKE DEMONSTRATION REACTOR PROJECT

J.W. HILBORN*, D.J. WINFIELD*
Atomic Energy of Canada Limited,
Ontario, Canada
Emails: hilbovanw@sympatico.ca, winfielddavid@bell.net

Abstract

In 1984 construction for the 2 MW(th) SLOWPOKE Demonstration Reactor (SDR) began at the Atomic Energy of Canada Limited Whiteshell Nuclear Research Establishment, Manitoba, to demonstrate the feasibility of low temperature district heating. First criticality was reached on 15 July 1987. The concept that SLOWPOKE type reactors could provide heating for communities, urban installations and for replacement of oil and diesel fuel in remote locations had by then been considered by Atomic Energy of Canada Limited for 20 years. The SDR reactor was successfully commissioned but was shut down after initial testing operations due to a combination of political, financial and technical reasons, before its full long term capability could be demonstrated. At that time the political climate was not conducive for further SDR operation nor for a proposed higher power scaled up version to be built. Today almost 40 years later proposals for multiple types of small modular reactors, intended for the same purposes that SDR was developed, are many and varied. It still remains to be demonstrated whether any of these currently proposed small modular reactors will be licensable, economically viable and technically successful in the long term, as was envisaged by the pioneering design of the SDR.
*Retired.

SLOWPOKE FACILITY AT POLYTECHNIQUE MONTRÉAL

C. CHILIAN, D.R. HALL, G.G. KENNEDY, J. ST-PIERRE
Department of Engineering Physics,
Polytechnique Montreal,
Montreal, Quebec, Canada
Email: cornelia.chilian@polymtl.ca

Abstract

The SLOWPOKE facility at Polytechnique Montréal is located at ground level in the six storey main building of the Polytechnique campus near downtown Montreal. It is one of four similar originally high enriched uranium fuelled SLOWPOKE reactors installed at Canadian universities in 1976. The facility has been extensively used for teaching nuclear engineering, for neutron activation analysis and to produce radioactive tracers. The reactor is still operating with the original analogue control system. In 1997 the reactor was the first SLOWPOKE to be converted from high enriched uranium to low enriched uranium fuel. Since 2010 the utilization of the fuel has been optimized, therefore the reactor is planned to continue operating beyond 2050. Between 2010 and 2022 the facility underwent major building infrastructure upgrades for electric power systems, asbestos removal, fire protection, digitalized security systems, centralized heating, ventilation and air conditioning. Since 2009 the two digital spectrometers were replaced and another two were added later, almost all irradiation and counting systems are nowadays digitalized, and a new automatic sample irradiation and counting system has been added. The pool water purification system has been upgraded in 2021.

CIAE MINIATURE NEUTRON SOURCE REACTOR

Y. LI., X. WU, Q. HAO,
M. WANG, S. ZOU, J. ZHANG
China Institute of Atomic Energy,
Beijing, China
Email: ygli@ciae.ac.cn

Abstract

The Prototype Miniature Neutron Source Reactor (hereinafter referred to as MNSR IAE) was designed and built by China Institute of Atomic Energy (CIAE) and reached first critical on 10 March 1984. The MNSR IAE rated thermal power is 27 kW. The reactor is based on a pool tank type design with a sealed reactor container vessel suspended and immersed in a light water pool. Heat from the core is removed by natural convection of light water moderator in the reactor. The reactor was originally fuelled with 90.3% enriched ^{235}U high enriched uranium U–Al alloy fuel. In 2016 the fuel was converted to use 12.5% enriched UO_2 fuel. Operated and maintained by the CIAE the MNSR IAE prototype has now currently been operating safely for 37 years. The main applications are neutron activation analysis, nuclear detector testing, personnel training, teaching and research in nuclear engineering and related topics.

SHANDONG MINIATURE NEUTRON SOURCE REACTOR

J. HONG, M. WANG, Q. HAO, D. OUYANG, Y. LI
China Institute of Atomic Energy,
Beijing, China
Email: hongjingyan@139.com

Abstract

The Shandong Miniature Neutron Source Reactor (MNSR-SD) had a rated power of 30 kW and was first critical in May 1989. The reactor was safely operated and maintained for 19 years by the Shandong Institute of Geological Sciences with no safety related events. Over the reactor's lifetime ten irradiation sites provided the means to analyse hundreds of thousands of irradiated samples for geological and environmental research, using neutron activation analysis. Other main utilizations were for neutron detector testing and calibration for power reactors, personnel training and teaching. Decommissioning began in 2008 and was completed in 2012. The reactor building was then authorized for future free use without restrictions.

SHANGHAI MINIATURE NEUTRON SOURCE REACTOR

D. PENG, Q. HAO, S.Y. ZOU, J.H. ZGANG, K.J. LI
China Institute of Atomic Energy,
Beijing, China
Email: pengdan.mnsr@139.com

Abstract

The 30 kW Shanghai Miniature Neutron Source Reactor (MNSR-SH) was designed by the China Institute of Atomic Energy and constructed during 1991 by the Shanghai Institute of Measurement and Testing Technology, who managed and operated the facility. First critical was reached on 15 December 1991, with full power in service operation in April 1992. MNSR-SH was an improved design of the prototype MNSR IAE. The main applications were neutron activation analysis, production of short lived isotopes, and training of technical personnel in nuclear applications and related nuclear research. The reactor was permanently shutdown in June 2004 after 13 years of incident free, safe operation. Decommissioning began in early 2006. Fuel unloading, fuel transportation off site and temporary waste storage activities were completed by the end of 2006. In 2008 the reactor site was approved to reopen, without restriction, for free use.

SHENZHEN UNIVERSITY MINIATURE NEUTRON SOURCE REACTOR

H. ZHAO, Q. LUO, S. HU
Joint Institute of Applied Nuclear Technology of Shenzhen University,
Shenzhen
Email: zhaohg@szu.edu.cn

D. OUYANG, K. LI
China Institute of Atomic Energy,
Beijing

China

Abstract

The 30 kW Shenzhen University Miniature Neutron Source Reactor (MNSR-SZ) was first critical on 1 November 1988. The MNSR-SZ is a pool tank type reactor, using light water as moderator and natural circulation coolant, beryllium as a reflector and U–Al alloy fuel with ^{235}U enrichment of 90.2%. MNSR-SZ is operated and maintained by the Joint Institute of Applied Nuclear Technology of Shenzhen University. It is mainly used for neutron activation analysis, analysis of trace and minor elements and biology and material irradiation experiments. Graduate and postgraduate teaching and experiments are conducted on this facility. It has been in safe operation for 35 years without any incidents.

GHANA RESEARCH REACTOR-1 FACILITY REPORT

H.C. ODOI, E.O. AMPONSAH-ABU,
H.K. OBENG, P. DORDOH-GASU,
I.K. BAIDOO, K. GYAMFI, W. OSEI-MENSAH, E.K. BOAFO,
E. SHITSI, R.E. QUAGRAINE, W.S. MASSIASTA
National Nuclear Research Institute,
Ghana Atomic Energy Commission,
Kwabenya, Accra, Ghana
Email: hencilod@gmail.com

Abstract

The Ghana Atomic Energy Commission (GAEC) is owner of the Ghana Research Reactor-1 (GHARR-1). The operating organization is the National Nuclear Research Institute, one of the seven institutions in GAEC. GHARR-1 is one of the Chinese Miniature Neutron Source Reactors (MNSRs), which are also operated in China, Ghana, the Islamic Republic of Iran, Nigeria, Pakistan and the Syrian Arab Republic. GHARR-1 has been in operation since December 1994 and now uses 13.0 % low enriched uranium oxide after it was converted from 90% high enriched uranium to low enriched uranium in 2017. It is used for research and development, education and training and human resource development. The purpose of the paper is to share operational and utilization experience with other facility operators of MNSR and SLOWPOKE type reactors.

ISFAHAN MINIATURE NEUTRON SOURCE REACTOR

M.H. CHOOPAN DASTJERDI, J. MOKHTARI
Reactor and Nuclear Safety Research School,
Nuclear Science and Technology Research Institute,
Atomic Energy Organization of Iran,
Tehran, Islamic Republic of Iran
Emails: mdastjerdi@aeoi.org.ir, jvmokhtari@aeoi.org.ir

Abstract

Isfahan Miniature Neutron Source Reactor (MNSR) is one of the four operational research reactors in the Islamic Republic of Iran that was put into operation in 1994. The reactor is owned and operated by the Nuclear Science and Technology Research Institute and is used for neutron activation analysis, neutron radiography, prompt gamma neutron activation analysis, irradiation testing of materials, training and education of students and personnel. Recently, some changes have been made to the reactor in order to expand the reactor utilization. Some external beam tubes have been added for out-of-core experiments and irradiations using neutron radiography and prompt gamma neutron activation analysis. A new computer control system has been designed and developed for easier training and operation, as well as to expand reactor physics experiments. Research and training programmes continue, along with the safe operation and maintenance of the reactor.

THE JAMAICAN SLOWPOKE-2 RESEARCH REACTOR

H.T. DENNIS, C.N. GRANT
International Centre for Environmental and Nuclear Sciences,
University of the West Indies,
Kingston, Jamaica
Emails: haile.dennis02@uwimona.edu.jm, charles.grant@uwimona.edu.jm

Abstract

The Jamaican SLOWPOKE-2 research reactor (JM-1) is a pool type reactor fuelled by approximately 6 kg of uranium enriched to 19.89% ^{235}U and has a nominal thermal power rating of 20 kW when operating at full power. JM-1 was commissioned in 1984 with 93% high enriched uranium fuel and was converted to low enriched uranium fuel in 2015. The reactor is located at the International Centre for Environmental and Nuclear Sciences and is owned and operated by the University of the West Indies, Mona Campus and is primarily used for neutron activation analysis of trace elements in studies related to health, the environment and agriculture as well as education and training.

FUEL CONVERSION OF THE ICENS SLOWPOKE-2 REACTOR AT THE UNIVERSITY OF THE WEST INDIES, JAMAICA

E. HAGBERG
Atomic Energy of Canada Limited,
Ontario, Canada
Email: hagberg@sympatico.ca

C. GRANT
International Centre for Environmental and Nuclear Sciences,
University of the West Indies,
Kingston, Jamaica
Email: charles.grant@uwimona.edu.jm

Abstract

A brief description of the SLOWPOKE-2 reactor and the benefits, as well as disadvantages, of its design for undertaking a fuel conversion is provided. The conversion of the ICENS reactor is then described in more detail.

THE NIGERIA RESEARCH REACTOR-1

S.A. JONAH
Centre for Energy Research and Training,
Ahmadu Bello University,
Zaria, Nigeria
Email: jonahsa2001@yahoo.com

Abstract

The Nigeria Research Reactor-1 (NIRR-1) is the first nuclear research reactor in the country. It was acquired as a training and educational tool towards the development of a more robust nuclear energy programme for Nigeria. In 1996, the research reactor project was initiated under a tripartite agreement between China, Nigeria and the IAEA via a project and supply agreement. NIRR-1 was commissioned in 2004 with a highly enriched uranium (HEU) core. The reactor was operated with the HEU core from 2004 to 2018. Under the aegis of global HEU minimization efforts and the Global Threat Reduction Initiative NIRR-1 conversion from HEU to low enriched uranium (LEU) began as an IAEA coordinated research project in 2006 that brought together all commercial and protype miniature neutron source reactor operators. The project was aimed at enhancing nuclear security in civilian facilities and reducing the availability of weapons grade uranium. The conversion of NIRR-1 has thereby enhanced nuclear security in Nigeria and globally. It has also provided fresh impetus for the extended utilization of NIRR-1 because of the additional operational lifetime now available with the LEU core.

PAKISTAN RESEARCH REACTOR-2

T. MAHMOOD, M. IQBAL
Pakistan Institute of Nuclear Science and Technology,
Islamabad, Pakistan
Emails: mtmalikpk@yahoo.com, masiqbal@hotmail.com

Abstract

The Pakistan Research Reactor-2 (PARR-2) has been operating since 1989. Pakistan Institute of Nuclear Science and Technology is responsible for safe operation and maintenance of this research facility. The core of PARR-2 consists of 344 fuel elements arranged within concentric rings in a fuel cage cylinder with a diameter of 230 mm and a height of 248 mm. The reactor core is located in the lower part of the reactor container vessel which is located in a subgrade pool structure of reinforced concrete. PARR-2 has a number of inherent safety features; one of the most important ones being a limited excess reactivity of 4 mk. There are ten irradiation sites for experimental purposes. PARR-2 is being utilized for neutron activation analysis, material properties studies, student experiments and training of personnel.

SYRIAN RESEARCH REACTOR

I. OTHMAN, M. BOUSH, A. SARHEEL
Atomic Energy Commission of Syria,
Damascus, Syrian Arab Republic
Emails: iothman@aec.org.sy, mboush@aec.org.sy, asarheel@aec.org.sy

Abstract

The Syrian Research Reactor-1 has been operated since 1996. The Atomic Energy Commission of Syria (AECS) is responsible for the safe operation and maintenance of the research facility. The management board of the AECS governs the reactor facility. The Chair of the Management Board is the Director General of the AECS. The operating organization of the Syrian miniature neutron source reactor (MNSR) is the Syrian MNSR Operating and Investment Division at the Department of Nuclear Engineering. The reactor is located in a vessel supported in an underground pool of a stainless steel lined and reinforced concrete structure. The fuel core is in the lower section of the reactor vessel and consists of 347 high enriched uranium fuel elements and 3 dummy elements, which form a fuel cage. The cage is located inside an annular beryllium reflector and rests on a lower beryllium reflector plate. The Syrian MNSR has a number of inherent safety features with a small excess reactivity of 4 mk. Five inner irradiation sites and five outer irradiation sites are available for experimental purpose. The main purposes of the Syrian MNSR are neutron activation analysis, production of some radioisotopes and education and training.

SURANAREE UNIVERSITY OF TECHNOLOGY RESEARCH REACTOR

C. KOBDAJ, R. NGONCHAIYAPHUM, S. TEPAROS,
S. JUNON, N. CHAWARATSINTORN, N. TANAPAKSUVAR
Boron Neutron Capture Therapy Research Centre,
Suranaree University of Technology,
Muang Nakhon Ratchasima, Thailand
Email: kobdaj@g.sut.ac.th

Abstract

The Suranaree University of Technology Research Reactor (SUT-RR) is a tank in pool natural convection cooled research reactor with a thermal power of 45 kW, using metallic beryllium as a reflector and UO_2 as fuel with a 19.75% ^{235}U enrichment and light water as coolant and moderator. The reactor core is located in a closed container vessel, which is suspended in a reactor pool. The main purpose of the reactor is for use of the boron neutron capture therapy technique. Other applications are research on neutron radiography, prompt gamma neutron activation analysis, neutron activation analysis, physics experiments, nuclear detector testing and teaching and training. SUT-RR is designed by the China Institute of Atomic Energy and is currently under ongoing construction. The Suranaree University of Technology will be responsible for the operation and maintenance of the reactor.

ABBREVIATIONS

ASDS	auxiliary shutdown system
BNCT	boron neutron capture therapy
CANDU	Canada Deuterium Uranium
CHF	critical heat flux
DCC	digital computer control
DNC	delayed neutron counting
HEU	high enriched uranium
LEU	low enriched uranium
MCNP	Monte Carlo N-Particle
MNSR	miniature neutron source reactor
NAA	neutron activation analysis
NEW	nuclear energy worker
OFI	onset of flow instability
OLCs	operational limits and conditions
ONB	onset of nucleate boiling
OSV	onset of significant void
PGA	prompt gamma ray analysis
RMC	Royal Military College
RRDB	Research Reactor Database (IAEA)
SAR	safety analysis report
SDR	SLOWPOKE heating demonstration reactor
SIRCIS	SLOWPOKE Integrated Reactor Control and Instrumentation System
SLOWPOKE	Safe Low Power Critical Experiment
SPFD	self-powered flux detector
SRC	Saskatchewan Research Council
TRIGA	Training, Research, Isotopes, General Atomics

CONTRIBUTORS TO DRAFTING AND REVIEW

Bennett, L.	Royal Military College of Canada, Canada
Boush, M.	Atomic Energy Commission of Syria, Syrian Arab Republic
Chakrov, P.	International Atomic Energy Agency
Chan, P.	Royal Military College of Canada, Canada
Chilian, C.	Polytechnique Montréal, Canada
Cols, H.	International Atomic Energy Agency
Dastjerdi, M.H.C.	Atomic Energy Organization of Iran, Islamic Republic of Iran
Dennis, H.	University of the West Indies, Jamaica
Dewes, J.	International Atomic Energy Agency
Duke, M.J.M.	University of Alberta, Canada
Farjallah, N.	International Atomic Energy Agency
Gavello, M.-K.	International Atomic Energy Agency
Geupel, S.	International Atomic Energy Agency
Grant, C.	University of the West Indies, Jamaica
Hao, Q.	China Institute of Atomic Energy, China
Hong, J.	China Institute of Atomic Energy, China
Iqbal, M.	Pakistan Institute of Nuclear Science and Technology, Pakistan
Janski, S.	International Atomic Energy Agency
Jonah, S.A.	Ahmadu Bello University, Nigeria
Kobdaj, C.	Suranaree University of Technology, Thailand
Laine, N.	International Atomic Energy Agency

Li, K.	China Institute of Atomic Energy, China
Li, Y.	China Institute of Atomic Energy, China
Luo, Q.	Shenzhen University, China
Malik, T.M.	Pakistan Institute of Nuclear Science and Technology, Pakistan
Mazzi, R.	International Atomic Energy Agency
Michal, V.	International Atomic Energy Agency
Mokhtari, J.	Atomic Energy Organization of Iran, Islamic Republic of Iran
Nielsen, K.	Royal Military College of Canada, Canada
Norrlid, L.D.R	International Atomic Energy Agency
Odoi, H.C.	Ghana Atomic Energy Commission, Ghana
Ouyang, D.	China Institute of Atomic Energy, China
Peng, D.	China Institute of Atomic Energy, China
Pessoa Barradas, N.	International Atomic Energy Agency
Ridikas, D.	International Atomic Energy Agency
Sarheel, A.	Atomic Energy Commission of Syria, Syrian Arab Republic
Shokr, A.	International Atomic Energy Agency
Stevens, J.G.	Argonne National Laboratory, United States of America
Sun, K.	International Atomic Energy Agency
Wang, M.	China Institute of Atomic Energy, China
Whitlock, J.	International Atomic Energy Agency
Winfield, D.J.	Consultant, Canada

Wu, X.	China Institute of Atomic Energy, China
Zhang, J.	China Institute of Atomic Energy, China
Zou, S.	China Institute of Atomic Energy, China

Consultants Meetings

Vienna, Austria, 27–29 April 2021, 5–8 September 2022

Technical Meeting

Vienna, Austria, 28–30 August 2024

# CONTACT IAEA PUBLISHING

Feedback on IAEA publications may be given via the on-line form available at:
www.iaea.org/publications/feedback

This form may also be used to report safety issues or environmental queries concerning IAEA publications.

Alternatively, contact IAEA Publishing:

Publishing Section
International Atomic Energy Agency
Vienna International Centre, PO Box 100, 1400 Vienna, Austria
Telephone: +43 1 2600 22529 or 22530
Email: sales.publications@iaea.org
www.iaea.org/publications

Priced and unpriced IAEA publications may be ordered directly from the IAEA.

ORDERING LOCALLY

Priced IAEA publications may be purchased from regional distributors and from major local booksellers.

Printed and bound by CPI Group (UK) Ltd, Croydon, CR0 4YY

06/07/2026

02160596-0003